C0-AJQ-902

Modern Telecommunication

Applications of Communications Theory
Series Editor: R. W. Lucky, *AT&T Bell Laboratories*

INTRODUCTION TO COMMUNICATION SCIENCE AND SYSTEMS
John R. Pierce and Edward C. Posner

OPTICAL FIBER TRANSMISSION SYSTEMS
Stewart D. Personick

TELECOMMUNICATIONS SWITCHING
J. Gordon Pearce

ERROR-CORRECTION CODING FOR DIGITAL COMMUNICATIONS
George C. Clark, Jr., and J. Bibb Cain

COMPUTER NETWORK ARCHITECTURES AND PROTOCOLS
Edited by Paul E. Green, Jr.

FUNDAMENTALS OF DIGITAL SWITCHING
Edited by John C. McDonald

DEEP SPACE TELECOMMUNICATIONS SYSTEMS ENGINEERING
Edited by Joseph H. Yuen

MODELING AND ANALYSIS OF COMPUTER COMMUNICATIONS NETWORKS
Jeremiah F. Hayes

MODERN TELECOMMUNICATION
E. Bryan Carne

Modern Telecommunication

E. Bryan Carne

GTE Laboratories Incorporated
Waltham, Massachusetts

Plenum Press • New York and London

Library of Congress Cataloging in Publication Data

Carne, E. Bryan, 1928–
Modern telecommunication.

(Applications of communications theory)
Bibliography: p.
Includes index.
1. Telecommunication. I. Title. II. Series.
TK5101.C2986 1984 621.38 84-16103
ISBN 0-306-41841-X

A Division of Plenum Publishing Corporation
233 Spring Street, New York, N.Y. 10013

Printed in the United States of America

Preface

Organized society depends on communication of all kinds, including the ability to communicate at a distance, instantaneously. With the development of solid-state electronics and its application to digital processing, telecommunication has become extremely important to large segments of American business. The introduction of competition to serve these voice, data, and video needs has expanded the number of service options available, and some of them are finding their way into the residential sector. From a relatively stable, mature industry, telecommunication has rapidly become a technology-driven marketplace in which a host of companies are competing for customer attention with new services and equipment.

Heretofore, books on telecommunications have addressed facilities and how they work. In this book, I am seeking to provide a much broader perspective which includes information on the motives driving the business itself, on new media and services, and on advancing technologies, as well as on digital facilities and their integration into the environment of future businesses and households. Covering so wide a set of topics presents many problems, not the least of which is that the character of the information is different in each chapter, and the material will be read by persons skilled in disparate fields. It is possible to read each chapter by itself—although a reading of all of them is needed to understand the new dimensions being introduced into the telecommunication experience. Where appropriate, I have referenced information in other chapters to encourage in-depth investigation, and have included a moderate level of references to other works for those who may wish to explore further. Because the field has so much specialized jargon, I have also included a glossary which contains the more important definitions and an explanation of acronyms and abbreviations.

In preparing this book I have been greatly helped by exposure to advanced telecommunication concepts in my work at GTE Laboratories Incorporated and by association with the extensive telecommunication services and equipment activities of GTE Corporation. I am grateful to my many friends in all parts of the corporation who have educated me in this field. Several of my colleagues were kind enough to review the first draft of this book. I am particularly grateful to Brian Dale, David Fellows, Dick Gordy, Ira Kohlberg, Jack Lawrence, and Moira Ounjian for their helpful remarks. The entire manuscript was prepared by Carol Oliver, who cheerfully coped with equipment failures and dataset problems. Her mastery of IBM Script and the associated machinery made revisions and rewrites almost fun. It is a pleasure to thank her for her support, and to thank Ellen Connor, who prepared the original illustrations. Finally, I should like to thank Paul Ritt, Vice President and Director of Research of GTE Laboratories Incorporated, for his friendship and encouragement over the many years during which the seeds for this book were sown and harvested.

E.B. Carne

Concord, Massachusetts

Acknowledgment

Chapters 2, 4, and 5 contain information extracted from works published previously by the Author. Specifically, Chapter 2 contains portions of "New Dimensions in Telecommunications" by E. Bryan Carne, *IEEE Communications Magazine,* January 1982 (pp. 17–25), copyright 1981 IEEE. Other information from this article appears in Chapters 4 and 5. In addition, Chapter 5 contains portions of "The Wired Household," by E. Bryan Carne, *IEEE Spectrum,* October 1979 (pp. 61–66), copyright IEEE 1979, and "Future Household Communications—Information Systems," by E. Bryan Carne, included as Chapter 6 in *Communication Technologies and Information Flow,* edited by Maxwell Lehman and Thomas J. Burke, Pergamon Press, Elmsford, NY, 1981 (pp. 71–86), copyright 1981 Pergamon Press, Inc.

Permission to include this material is gratefully acknowledged.

Contents

Introduction

In recent years innovative services have come to characterize the modern telecommunications industry. The pressure for these new services comes partly from customer demand and partly from the availability of new technology. Customer demand arises from the needs (real or imagined) for

- rapid response—for immediate access to information from, or the completion of transactions with, a remote data base;
- immediate communication with geographically dispersed units of an organization, or relatives and friends;
- circumvention of inconvenient opening hours through shopping by telephone, or automated tellers, or bill-paying services;
- remote interrogation and control of security and other environmental systems;
- verification of credit and authorization of the use of funds;
- nationwide television broadcasting and new television formats such as high-definition, expanded screen services;
- message communication in a highly mobile society, such as electronic mail, high-speed facsimile, and voice storage.

Each of these examples is supported by special interest groups. Some have a genuine requirement which is turning into market pull and causing service evolution and growth. Others have a technology able to provide more advanced capabilities and are looking for markets. Technologies which are pushing for applications are

- very large scale integrated (VLSI) circuits which provide more memory and processing power than has been available before on a single chip;
- digital switching and transmission which provide easier telecommunication between computers, data processing machines, and terminals, and promise cheaper, more reliable services for all;

- multifunction work stations with easy to use features;
- bandwidth reduction devices which make digital voice and digital video easier to handle;
- optical fiber transmission systems which offer high-capacity, interference-free transmission pathways;
- geostationary satellite systems which facilitate communication between points irrespective of terrestrial constraints, and distribute wideband signals over one-third of the surface of the earth.

Each of these examples is supporting advanced services of some sort. It is the full potential of these technologies that the technical community is seeking to exploit. Between technologists and users is a third group—entrepreneurs ready to match technology and market so as to create a business opportunity. Their job is perhaps the hardest of all for they must balance the enthusiasm of technologists against the fickleness of users, competition with a better idea, and the doubts of financiers. In addition, they operate in a newly competitive arena in which the rules are not totally clear and may be changed by administrative, legislative, and judicial fiat.

It is this modern telecommunication environment that is addressed in this book. The subject matter is divided into five chapters. In Chapter 1, telecommunication is defined, a short historical perspective is given, user demands and supplier motivations are explored, and the changes in the telecommunication industry due to FCC initiatives and an antitrust settlement are described. Chapter 2 is devoted to a description of the new electronic media, their uses and limitations, and a discussion of user reactions. Based on existing record and real-time media, they provide opportunities to retrieve information, to exchange messages, and to conference to various degrees. The possibility of other electronic media is explored. Chapter 3 provides a qualitative review of pertinent engineering science. It includes sections devoted to voice, data, and video signals; elements of transmission and digital switching; traffic theory, software, and protocols; artificial intelligence; and solid-state electronic and optical devices. These topics are the essential underpinnings to an understanding of the functioning of modern facilities. Chapter 4 describes a range of primarily digital facilities which support telephony, television and radio, and data communication. Telephony facilities include interexchange and exchange area transmission, switching and networks, and telephones and private branch exchanges. Television and radio facilities include advanced television receivers, cable television systems, and direct broadcast satellite systems. Data communication facilities include data terminals, data switching and transmission, packet networks, and local area networks. In Chapter 5, media, technology, and facilities are combined in a discussion of future network, switching, and transmission arrangements which will provide productivity improving services for business and personally satisfying services at home.

To some degree, each chapter stands by itself, and can be read alone. Most readers may prefer to start with Chapter 1, proceed to either "Media," "Technology," or "Facilities," and then branch out from there, ending at Chapter 5. Others may find reading Chapters 1 and 5 will direct them to topics in the central three chapters. Inevitably, the discussion of "Facilities" (Chapter 4) depends on "Technology" (Chapter 3), and "Integration" (Chapter 5) depends both on "Media" (Chapter 2) and "Facilities" (Chapter 4). Where important, note is made of associated sections in other chapters. Where appropriate, references to recent literature are included, although no attempt has been made to provide an exhaustive bibliography. The Glossary contains definitions of technical terms and the meaning of acronyms and abbreviations.

1

Background

In common with other segments of the contemporary environment, telephony, radio, and television have benefited from the introduction of solid-state devices and digital computer techniques. In addition, in the United States, the measured pace of telecommunication evolution in a regulated market, which characterized the first three-quarters of this century, has been shattered by the adoption of a procompetition philosophy by regulators, the breakup of the world's largest telephone company under the supervision of a federal court, and the entry of new firms into the telecommunication market that possess substantial resources earned from other endeavors. The result is a highly fluid environment in which providers seek to interest users in a plethora of advanced telecommunication services and computer-based equipment which have applications at work and at home.

1.1. What is Telecommunication?

Literally, *telecommunication* (*tele*—from the Greek τηλε—meaning "far off") is *communication at a distance*. Communication is generally understood to be the act of imparting or exchanging information, knowledge, and ideas. Originally, it required people to come together to speak to one other. Later, with the invention of writing, messages were carried from place to place, so that communication could occur at a distance, but only as far, and as fast, as a person could travel. Improvements in transportation resulted in improvements in the speeds and distances at which communication occurred, making the organization of states, enterprises, and communities of ideas possible. With the introduction of the telegraph and the telephone, limitations of both distance and time were removed, and instantaneous communication over any distance is now available to most of the world's population. Moreover, the same technology which resulted in electronic communication has given rise to computers which are programmed to exchange data with other machines in other locations, and provide information to persons through the mediation of terminals.

The concept of communicating at a distance is shown three ways in Figure 1.1: as information transfer between sending and receiving parties or devices, as a connection of facilities, and as a particular service supported by available media. Telecommunication requires *facilities* which collect, transmit, and deliver information by electrical, electronic, and optical means. Such facilities include terminals, links, and nodes arranged in networks which can contrive whatever connections are necessary to transport messages between the sending parties or devices and the receiving parties or devices. They provide capabilities which support the development of telecommunication *media* and enable telecommunication *services* to be provided.

Media are agencies or means which facilitate actions at a distance. They may be divided into *record* media, which present a limited amount of information in a finite time, and preserve it for study; and *real-time* media, which deliver a continuous stream of information as it is produced, but preserve none of it for study and future reference. Thus, books are a record medium—they store information and facilitate the transfer of ideas from the writers to the readers. In contrast, television is a real-time medium which transports the sights and sounds of far-off activities to the viewer but preserves none of them. Of course, segments of the information transported by real-time media can be preserved in record media such as video or audio recordings, or can be the basis of a description contained in a book.

Telecommunication media are differentiated by their *contents,* which are related to the types of messages transported (print, visual, audio, data, etc.) and

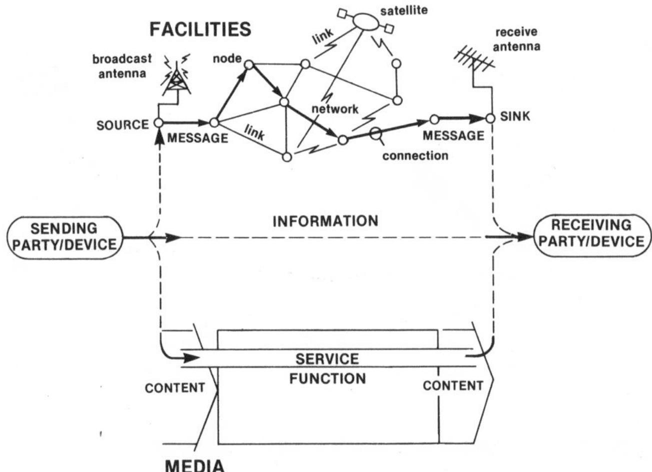

Figure 1.1. Telecommunication is the act of communicating instantaneously at a distance. The process may be characterized as information transfer and described in terms of facilities or media.

the *functions* they perform, which are related to the degree of interaction they support between sending and receiving parties (one-way broadcast, two-way point-to-point, etc.). Telecommunication services are delivered by media. Each represents a subset of the contents and functions of the medium tailored to satisfy a particular need. It follows that, to provide a specific service requires a medium whose contents and functions are at least equal to those which will be required to handle the types of messages and to provide the degrees of interaction needed to deliver the service.

Telecommunication services can be classified as *intercommunication* services which allow the exchange of voice, data, or video messages among two parties, or a small number of parties in conference; and *mass communication* services which provide voice, data, or video messages to any number of receiving parties, usually in a given area. Telecommunication services may also be divided into *business* services, which reflect the needs of industry and commerce, and *residence* services, which reflect the needs of the household. Predominantly, business services use media which can be supported by telephone, and telephone-compatible equipment, connected through networks established by common carriers, and over customer-owned facilities, or a combination of them. Almost exclusively, they are intercommunication services. In the home, the dominant services are those supplied by radio and television. Other services are supplied over the public switched telephone network. Residence services, then, are primarily mass-communication services.

1.2. Historical Perspective

1.2.1. 801st Lifetime

In *Future Shock,* Alvin Toffler[1] pointed out that *Homo sapiens* has existed for the equivalent of some 800 modern lifetimes (approximately 50,000 years). Throughout most of this time, spontaneous communication has only been possible for those within earshot. For less than the last 80 lifetimes have we been able to write; for only the last two lifetimes have we had the ability to communicate instantly, at a distance; and for less than a single lifetime have the sound and sight of worldwide events been available to large segments of the population, simultaneously, as they happen. Against the perspective of human history, telecommunication is a very recent occurrence. If the present pace of development is maintained, the 801st lifetime will be interesting, indeed!

1.2.2. Telegraph, Telephone, Radio, and Television

In the early years of the 19th century, Oersted, Henry, Gauss, and Weber, among others, developed the sciences of electricity and magnetism. In the 1840s the work of Samuel Morse and Alfred Vail turned theory and experiment into practice to produce the *electric telegraph*. Using it, trained operators could exchange coded messages with one another over distances limited mostly by the availability of the copper wires which were quickly strung over the countryside. In the 1870s, the invention, and in the 1880s, the exploitation, of the *telephone* by Alexander Graham Bell, Elisha Gray, and others, allowed ordinary people to talk to one another at ever increasing distances. Soon, under the leadership of Theodore Vail, cities were being buried under strands of wire strung on taller and taller poles, as the *telephone network* took shape. By the beginning of the 20th century, Guglielmo Marconi had demonstrated the principle of *wireless* communications, and wireless telegraphy found application to telecomunication with ships at sea. In the 1920s, a proliferation of *radio* stations brought entertainment, news, and eyewitness descriptions of far-off events to everyone who wished to listen. In the 1940s, sight was added to sound—*television* was born. For many, it opened a window on the United States and the world for the first time. In the 1950s, for those who lived in poor reception areas far from urban centers, the window quickly became a community affair delivered by *cable* connected to a common antenna sited on a convenient hilltop. In the 1960s, the window was filled with living color. Shortly thereafter intercontinental satellites made it possible to view events in real time across the oceans, and the space program extended our vision to the moon and planets.

1.2.3. Data Communication

Sound and sight were not the only messages which modern society demanded. In the 1950s, the *computer* emerged from universities and research institutes into the realms of commerce and industry. As well as being a tool to perform complex calculations, it was perceived to have the added promise of *data processing*. By the 1960s, reliance on these machines for more and more commercial functions produced a need to connect remote terminals to central processors, and processors needed to exchange information, giving rise to *data communication*. Digital signals were converted to analog signals to make use of the capabilities of the telephone network. At the same time, in order to determine better ways to transmit data, experiments began which recognized the short-term, bursty nature of much of the traffic. Eventually, the concept of *packets* was proved and *packet networks* were established.

1.2.4. The Information Society

The beginning of what many call the *information society* has stimulated a multitude of new facilities and specialized networks. In the 1970s, continuing advances in computer, communication, and semiconductor technologies encouraged new forms of services such as *electronic mail,* television-based information retrieval services known as *teletext* and *viewdata,* and *audio, video,* and *computer conferencing*. In turn, these developments have produced concepts for *automated offices* and *wired households*—models of business and residential telecommunication which can provide a rich combination of services expected to be of use in future offices and households. Many of these services will be supported by the technologies and facilities described later.

Figure 1.2 places the development of telecommunications in perspective with the development of human skills. Only a logarithmic time scale makes it possible to detail the events of the last 100 years. The 801st lifetime and the full development of the information society will occur on the right-hand extension of the time line. Figure 1.3 shows the buildup of these services since the early 19th century, and suggests their amalgamation into the intelligent network of the 21st century which will furnish services to support the sophisticated environment of a postindustrial society built upon voice, data, and video messages.

1.2.5. Distribution of Resources

Widespread telecommunication is a 20th century phenomenon. Facilities valued at several hundred billion dollars are in place. They are operated by a

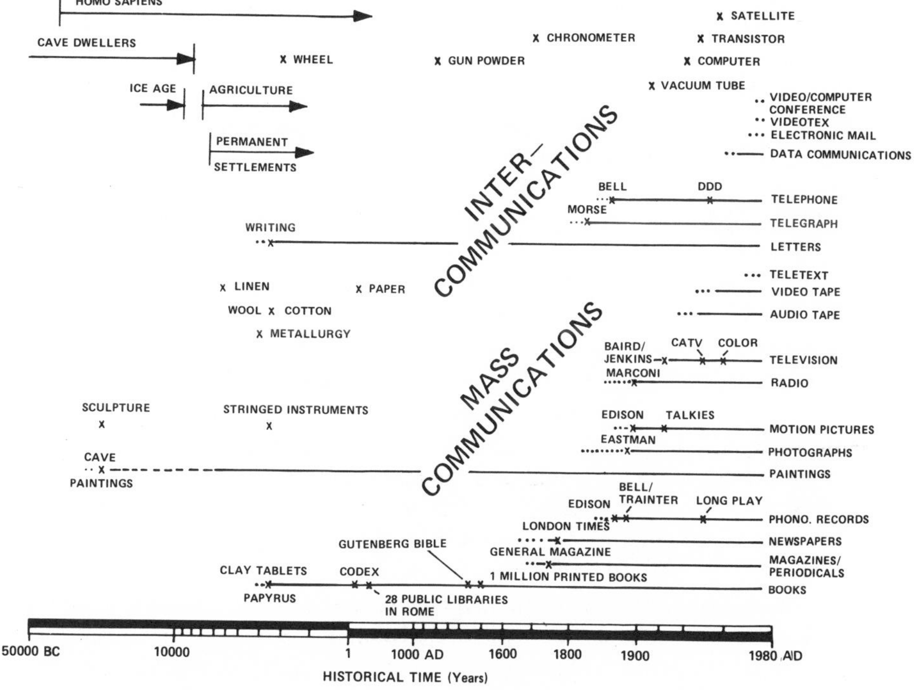

Figure 1.2. Perspective of the development of telecommunication facilities, media, and services against the development of human history. Only a logarithmic time scale allows the events of the last 100 years to be seen.

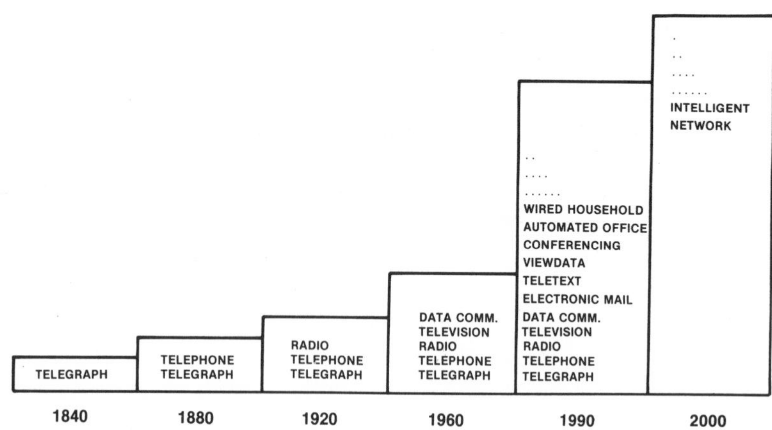

Figure 1.3. The buildup of telecommunication media and services from 1840 to 2000.

penditures for replacements, extensions, and new installations are approaching one hundred billion dollars. So large a market engages the attention of the world's major corporations, not to mention a host of smaller organizations. In common with many other resources, the ability to communicate instantaneously, at a distance, is not evenly divided among the population of the world. In round numbers, there are nearly one-billion radio receivers, 500 million television receivers and 500 million telephone instruments in use. Fully five-sixths of them are located in the developed nations, and one-half of this number are to be found in North America. On average, persons in North America have nearly 20 times as many telecommunication terminals available to them as persons in the third world, and three times as many terminals available to them as persons in the other developed nations. Truly, the North American continent is telecommunication rich!

1.2.6. Service Providers

In the United States, a federal policy of encouraging competition in the provision of telecommunication services is fueling growth and diversification. The result is a proliferation of providers, and an emphasis on new services for which there is a demand, and from which there is the likelihood of profit. In other developed countries, central telecommunication authorities are responsible for the development and exploitation of advanced telecommunication capabilities. For them, the social benefits of new services are at least as important as the commercial benefits. In some of these countries the central control is being loosened to allow limited competition. In most of the rest of the world, telecommunication is a tightly controlled function of the central government and competition to provide telecommunication services is unlikely.

1.3. Demand and Motivation

1.3.1. User Needs and Supplier Rewards

New products and services develop through the interaction of user needs, which produce demand; supplier rewards, which produce entrepreneurial motivation; and the availability of suitable technology. For products and services which use available technology and are designed to fill an existing need, demand can be assessed, production costs can be calculated, and profit can be projected. With this information, the probability of a successful outcome can be estimated and the attractiveness of the market (market *pull*) to any particular supplier can be determined. The degree of market pull tempered by the degree of risk the potential supplier will accept provides a logical basis for a decision on whether to proceed with development and introduction. However, if the products and services are designed to demonstrate the application of new technology in the hope of stimulating demand (technology *push*), it is impossible to project these factors with any certainty. Statistical evidence suggests that innovation to fill a need is several times more successful than innovation to exploit a technical development.(2)

1.3.2. Market Pull

1.3.2.1. Business Applications

For organizations operating in economies in which market shares and product or service margins are established by competition, there is a continuing need to function more efficiently—to improve productivity. In combination with computers and electronic data banks, telecommunication can instantly transport information to wherever it is required and return instructions and new information to be processed and acted on. Insofar as this activity can replace people or speed up operations, telecommunication results in improved productivity. Insofar as telecommunication makes other functions possible, and increases the amount of information handled, it increases the sophistication of the operation and may provide opportunities both for additional savings and for increased output. For any operating improvement which can be shown to have an expectation of a reasonable return on investment, there is an innovator ready to finance and introduce it, and for changes which demonstrate a continuing return, there is a market. Thus new telecommunication media which contribute positively to productivity improvement will be adopted. In the United States, facilities which support data processing, information retrieval, electronic message distribution, and teleconferencing are being offered by both existing and new telecommunication suppliers, and are being incorporated into the fabric of the business sector. In other nations, their introduction is proceeding more slowly. Data communication and associated activities are the prime example of new telecommunication capabilities shaped by market pull.

1.3.2.2. Residence Applications

Residence needs are more difficult to state than business needs because they include psychological factors which provide powerful motivations whose priorities change with time and personal circumstances. Unlike business, economic advantage may not be the single most important criterion by which to evaluate new residence opportunities. Personal gratification, enhancing self-respect, raising mutual esteem, reinforcing a sense of community, and similar factors, can be of overwhelming importance. Focus on self, and stress on leisure can make the least-cost solution quite unacceptable. Under these circumstances there may not be sufficient common ground among groups of people to provide a market large enough for a specific new product or service to achieve an acceptable price through scale-economies (see Section 1.3.4.1). Fire alarms, burglar alarms, medical alerts, and companion services for the elderly, are important for segments of the residential community. Resource conservation (if it saves money), education (if it is relevant), and information retrieval (if it is convenient) may also receive support. At present, there is no market pull encouraging the large-scale introduction of new telecommunication products and services into the home.

1.3.3. Technology Push

1.3.3.1. Effects of New Technology

By making fresh opportunities available, new technologies can affect the development of telecommunication facilities and services as certainly as the needs of society. Thus, the availability of advanced technologies permits the development of new media which support new services that may satisfy an existing need, and perhaps replace existing media. Alternatively, the new medium may uncover a latent need, thereby providing a new service. These are the *direct* effects of new technology.

The application of advanced technologies may lead to changes in the performance of existing facilities and the cost of existing services, thereby stimulating new operating strategies and network concepts. Some of these may be reflected in increased demands on existing media: others may make new facility configurations cost-effective, leading to new or improved services. These are *secondary* effects of new technology.

The existence of advanced technologies stimulates the development of new products and services in other areas. Some of these may eventually require communications capabilities which place new demands on existing media, or requirements for new media may be developed. These are *indirect* effects of new technology.

1.3.3.2. Forcing Technologies

It is generally accepted that an overwhelming majority of all of the technologists who have ever lived are alive today. They form a larger than critical mass which is fueling a technology engine the likes of which the world has never seen. Some of the changes which have occurred in technical parameters in telecommunication-related fields since 1970, and projections for the future, are shown in Figure 1.4. The vertical axis is a logarithmic scale on which each unit represents a multiplication by ten. Since 1970, the quality of optical fibers, as measured by attenuation, has improved 100-fold. This has resulted in up to a 100-fold increase in the distance that information can be transmitted before it

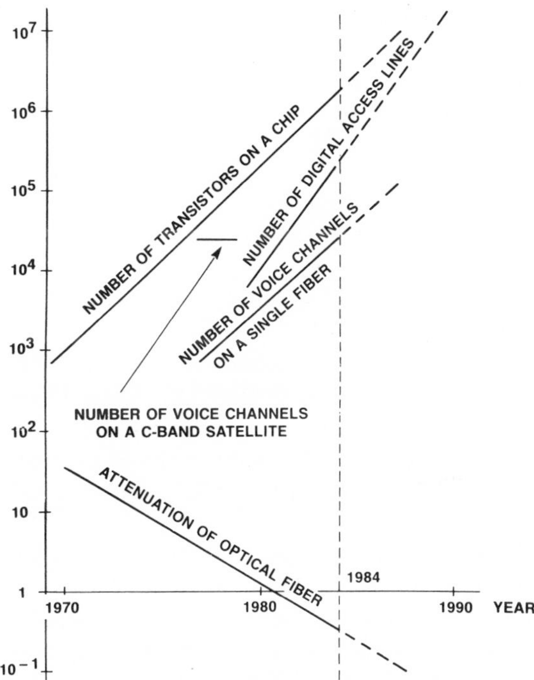

Figure 1.4. Changes which have occurred in some technical parameters of importance to telecommunication over the last 15 years.

must be amplified. The number of transistors which can be fabricated on an integrated circuit chip has increased by more than 1000-fold, paving the way for computers and other *systems-on-a-chip*. The number of digital access lines in service is increasing at a similar rate. The numbers of voice circuits which are carried on a single fiber are equal to the total number of voice circuits carried over a satellite, and there have been many changes of lesser consequence. The result has been to provide more options for telecommunication facilities and services than the traditional suppliers can accommodate. In the United States, this has led to the rise of new telecommunication organizations that offer advanced equipment and services in competition with established providers.

1.3.4. Effect of In-Place Facilities

To be the only person possessing a telephone is to have something that is useless. At the very least, telecommunication requires two parties with a common interest and compatible equipment. What is better is to have most people served by equipment which can connect them in whatever combinations are appropriate to the moment, or can receive and reproduce whatever signals are being broadcast at the moment. In order to achieve the full potential of worldwide communication, commonality between equipment is essential. In recognition of this, the nations of the world have formed committees and organizations to recommend standards, procedures, and practices for the common good. Because it is to everyone's advantage, telecommunication is an area of global activity in which cooperation is the rule.

1.3.4.1. Economies of Scale

Standardization has important economic advantages. In particular, in the telephone network, moderate to high economies of scale can be achieved by operating a common system through which an increasing number of messages flow. Lower, but still significant, economies of scale can be achieved through the increased manufacturing volumes produced by common equipment.[3] Thus, low-cost telecommunication is best fostered by minimizing the number of different systems so as to maximize the traffic each carries, and minimizing the number of different types of equipment so as to maximize the manufacturing volume of each type.

For many years, the world's major telephone operating organizations have striven to preserve this cost-minimizing strategy, with the result that they have been slow to incorporate new technology and provide capabilities to support new requirements. Further, the financial structure of the operating companies reflected a long-term view of equipment life, and the tariffs under which they functioned only allowed for slow rates of capital recovery. Replacing equipment on an actuarial basis resulted in a slow, but orderly, upgrading of capabilities and predictable capital requirements. In today's surging technical environment where new products compete to replace existing products long before installed equipment has reached the end of economic life, where improved performance comes at lower cost, and where new capabilities stimulate demands for new services, these policies conflicted with the opportunity to add new dimensions to the telecommunication experience. Within the last few years, the FCC and a number of state regulatory agencies have recognized today's faster rate of technological progress and our inflation-driven economy. Shorter depreciation periods are being introduced which have a favorable impact on the rate of capital recovery.

1.3.4.2. Response to New Demand

When presented with new demand, established telecommunication services providers can respond in four ways: ignore it, expand existing facilities, overlay existing facilities with new equipment, or replace existing facilities with new equipment. The advisability of ignoring demand depends very much on the political realities surrounding the provision of telecommunication services. In the United States, lack of response from the existing carriers to demands for data communication led to opening up telecommunication to competition. The result has been a proliferation of telecommunication providers, concentrating on data communication and other needs, who have overlayed existing facilities to provide United States commerce with services which lead the world.

Expanding existing facilities to cope with new demands is the action telecommunication providers take best: economies of scale can be reinforced, and the use of present equipment can be extended, or more cost-efficient equipment, perhaps using new technology, can be installed. If what is needed is more of the same services, or even new services which use an existing class of equipment, the present service provider is well able to fill the need. But what if the demand is not for present or related services? What if the requirement is for a stereo AM radio broadcast service, or a high-resolution television service—or data communications? Under these circumstances the established providers cannot simply expand existing facilities, but must install new equipment—something which can be done by others if they are permitted to supply the market.

In some cases, new companies have built entire facilities which overlay existing facilities and are operated independently of them. In other cases, new companies have built partial sets of facilities which connect to facilities leased from the established providers to complete their network. Examples of these arrangements are given in Figure 1.5. To the user, the effect is the same as a new network. To the service provider, the arrangement offers reduced investment, and the opportunity to take advantage of scale economies where possible. To the existing provider, the arrangement offers increased revenue from certain connections—albeit, perhaps, at the expense of losing revenue in other parts of the network. For this reason the existing provider may seek to satisfy the demand

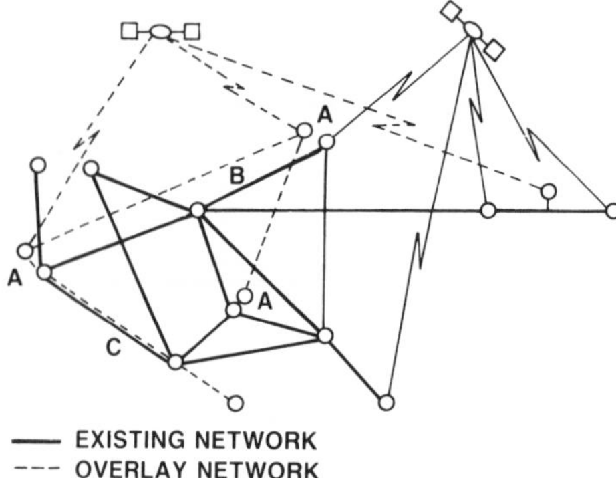

Figure 1.5. A network of existing facilities (solid lines) which is overlayed by a network of new equipment (dashed lines). At points marked A, the two networks have nodes which serve the same area. At B the networks have parallel, but separate facilities. At C the overlayed network uses facilities of the existing network.

by adding unique equipment where absolutely necessary and promoting maximum synergy between in-place and new facilities. Since the mid-1960s, this area of activity has been a battleground for entrepreneurs, regulators, legislators, and established service providers seeking an equitable division between *basic* and *enhanced* services (see Section 1.4.2) for voice, data, and video applications.

The fourth option to satisfy demands for new services—replacing existing facilities with new equipment—is available only to established providers. As noted earlier, past policy provided for an orderly replacement and upgrading of equipment. But what if demand cannot be satisfied by the existing equipment, the equipment is not scheduled for replacement, and there is new equipment which can perform the present functions, and those required to meet the new demands? The established provider can

- retire the in-place equipment, install the new, and seek higher tariffs;
- retain the in-place equipment and accept loss of revenue to a new company which can provide the new investment more readily;
- seek to augment the in-place facility with new equipment until the in-place equipment can be retired without penalty.

In practice, the last option is difficult to achieve on any large scale, and would most likely be noncompetitive unless major economies resulted.[(4)]

In Figure 1.6, a simple model is used to illustrate the last two options. In diagram A, the demand for service I is assumed to be growing at a steady rate. It is satisfied by the orderly extension and expansion of existing facilities. At a certain time ($t = 0$) demand begins to grow for service II which cannot be met by existing facilities: however, new equipment is available which can satisfy both services. In diagram B, the existing supplier stops expanding existing facilities at $t = 0$ and begins to install new equipment. As existing facilities are fully depreciated, they are retired and replaced. After a while the demand for both services is handled entirely by new equipment. In diagram C, the existing

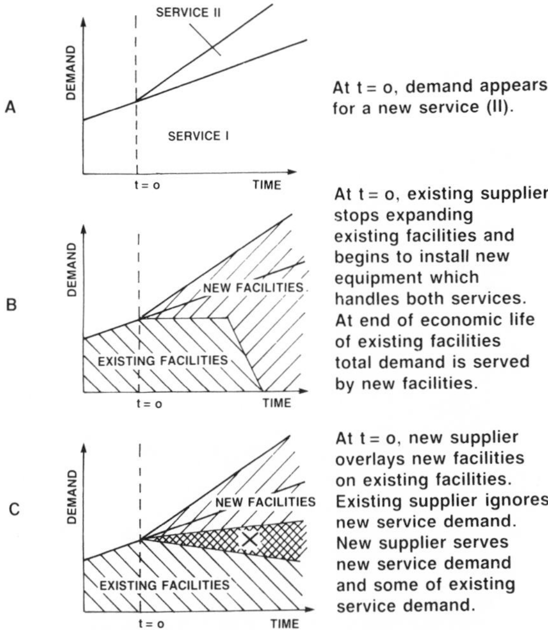

Figure 1.6. Ideal response to new service demand from an existing supplier, and a new supplier overlaying an existing supplier who ignores new demand. The area × in C represents demand which can go to either supplier depending on aggressive pursuit of the market.

supplier ignores the demand for service II. It is satisfied by a new supplier who overlays new facilities on existing facilities and proceeds to serve those users requiring service II and some of the demand for service I. The area marked *X* represents demand which can go to either supplier depending on how aggressively each pursues the market.

1.4. Changes in the Telecommunication Industry

1.4.1. Deregulation

For many years the implementation of new facilities to provide advanced services was slowed by government regulations which vested monopoly powers in the established providers of services. In the late 1960s, the idea developed that technology and the marketplace could create facilities which would provide the advanced services being demanded by commerce and industry, and *competition* became the watchword of the regulators. Throughout the 1970s, intelligent maneuvering by the owners of in-place facilities kept deregulation contained. Not until the early 1980s was it possible to perceive significant progress and to expect that market forces might be allowed to shape future telecommunication capabilities in the United States.

1.4.2. FCC Initiatives

Much credit goes to the Federal Communications Commission (FCC), which continued to assert procompetition policy in the face of what at times must have seemed to be overwhelming opposition. In a succession of orders in the late 1960s and the 1970s, the FCC

- established the principle that *foreign* equipment (i.e., equipment not supplied by the serving company) could be attached to the public switched telephone network;
- recognized categories of traffic which can be carried by *specialized* carriers (i.e., carriers seeking to serve specialized segments of the telecommunication market, as opposed to common carriers, which serve all segments);
- authorized *value-added* carriers (i.e., carriers which employ existing transport facilities and process the customer information in some way);
- developed the principle of *intercity* carriers (other common carriers—OCCs) competing for long-distance voice and data traffic;
- distinguished between *basic* services, which may, and *enhanced* services, which may not, be provided by a *dominant* common carrier (i.e., a carrier with enough market share to dominate the market);
- separated the provision of customer premise equipment (i.e., telephones, PBXs, etc.) from the provision of basic services by a dominant common carrier.

1.4.2.1. Basic Services

Basic services are transport services used for the collection and delivery of traffic on the behalf of others. The information delivered matches the information collected. Although the signal may have been changed in various ways on its journey through the network in order to facilitate passage from one equipment to another, the message content is unaffected. Signal processing associated with basic services is said to be *transparent* to the user, that is, the user is unaware it has occurred.

1.4.2.2. Enhanced Services

Enhanced services involve processing the message at some node or termination of the network in response to its content. Thus, electronic mail is an enhanced service in which a message is stored until delivery is requested by the person to whom it is addressed.

1.4.3. Antitrust Settlement

The keystone, for the restructuring of the telecommunication industry, was set in place by the settlement of an antitrust suit brought by the Department of Justice (DOJ) against American Telephone and Telegraph Company (AT&T) based on the contention that AT&T had used its control of local exchange facilities to frustrate competition from those wishing to be alternative suppliers of telecommunication services. Known as the Modified Final Judgement (MFJ), the agreement resulted in the separation of the *natural monopoly* of local exchange facilities and services from the *competitive* areas of interexchange switching and transmission, and customer premise equipment.(5) The result has been the divestiture of the *Bell operating companies* (BOCs) by AT&T and the creation of seven independent *regional operating companies* (ROCs) made up of combinations of BOCs. They provide basic services over exchange area facilities owned by the divested BOCs. Facilities which provide interexchange (i.e., long-distance) switching and transmission and furnish customer premise equipment have been consolidated into a new AT&T which is free to offer enhanced services in competition with all comers.

The BOCs are responsible for the operation of *exchange* areas which encompass one or more contiguous end offices and serve a community with common social, economic, and other interests. With a few exceptions, exchange areas do not bridge *standard metropolitan statistical areas* (SMSAs) nor do they cross state boundaries. They are contained in *local access* and *transport areas* (LATAs) which define calling areas (such as the area represented by an *area code*). Traffic within each exchange area (intraexchange) is carried on facilities owned by the BOC. Interexchange area, intra-LATA traffic may be carried over BOC facilities, non-BOC exchange carrier facilities, or facilities provided by other common carriers (OCCs). Interexchange area, inter-LATA traffic is carried on AT&T interexchange (ATTIX) carrier facilities or OCC facilities. These types of traffic are illustrated in Figure 1.7. In this way the regulated exchange area is separated from the competitive long-distance environment.

In the United States approximately 80% of all access lines are owned by BOCs. The other 22 million lines belong to some 1400 independent operating companies.* While the MFJ approved an agreement between AT&T and the DOJ, *de facto* it affects the entire industry. In particular, it separates long-distance and local revenues and foretells the end of the practice of one subsidizing the other. Stopping the revenue flow from long distance to local has resulted in the addition of an *access charge* paid by long-distance carriers for the opportunity to connect through exchange area facilities to individual customers, and by

*The exact number of independent operating companies is difficult to ascertain. It is slowly declining due to amalgamations and acquisitions. New technology can be expected to accelerate this activity.

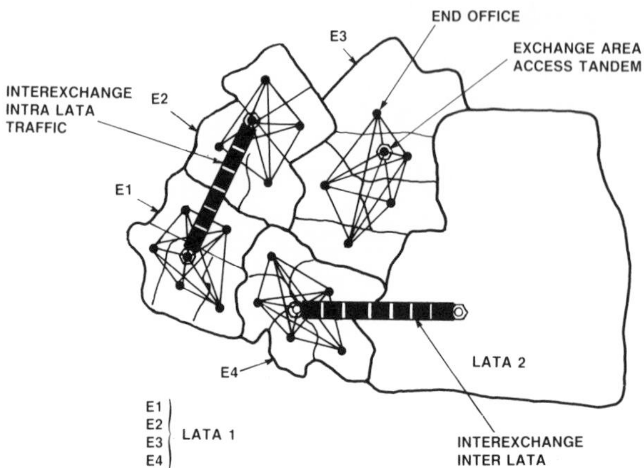

Figure 1.7 Divisions established by the MFJ. Exchange areas E1, E2, E3, and E4 make up LATA 1. Interexchange intra-LATA traffic may be carried over regional operating company facilities or by other common carriers (OCCs). Interexchange inter-LATA traffic may be carried over ATTIX or OCC facilities.

multiline business customers for the opportunity to connect through exchange area facilities to long-distance carriers. One day, similar charges may be levied on residential and single-line business customers.

1.4.4. Other Actions of the FCC

While the MFJ was the most significant event fostering the evolution of the telephone industry to a new competitive structure, other actions of the FCC have also had their effect:

- authorization of cellular mobile radio (CMR) systems in many metropolitan areas in the United States has added new competition between telephone companies (wireline carriers), independent radio carriers (IRCs), and others;
- authorization of digital termination service (DTS) and the integration of local and long-distance facilities to form digital electronic message service networks (DEMSNETs) has encouraged the development of technology which can *bypass* the local exchange facilities;
- authorization of direct broadcasting satellites (DBSs) has made it possible for truly national television broadcasting in competition with the established broadcasting networks and CATV systems.

At the same time, a court decision has made distant signal importation more costly for CATV system operators by asserting that copyright fees must be paid for the use of these signals. As never before, the telecommunication industry is being changed by administrative and judicial fiats. The result is a diversity of delivery systems employing advanced technology which seek to provide an expanding spectrum of services.

1.4.5. The New Environment

The greatest immediate effect of these activities was the breakup of the largest telephone organization in the world. Effective January 1, 1984, AT&T and the Bell System was split into eight separate organizations:

- *American Telephone and Telegraph.* (Estimated 1984 revenues, $56.6 billion; assets, $34.3 billion.) The corporation is divided in two:
  AT&T Communications. Domestic and overseas long distance carrier. (Estimated 1984 revenues, $37 billion.)
  AT&T Technologies. Includes AT&T Bell Laboratories (research and development), AT&T Network Systems (telecom equipment for carriers), AT&T Information Systems (equipment and business systems), AT&T International (foreign sales), and AT&T Consumer Products (telecom equipment for customers).
- *Ameritech.* Local telephone service in Wisconsin, Illinois, Indiana, Michigan and Ohio. (Estimated 1984 revenues, $8.3 billion; assets $16.3 billion; access lines, 14 million.)
- *Pacific Telesis.* Local telephone service in California and Nevada. (Estimated 1984 revenues, $8.1 billion; assets, $16.2 billion; access lines, 10.9 million.)
- *Bell South.* Local telephone service in Kentucky, Tennessee, Mississippi, Louisiana, Alabama, Florida, Georgia, North and South Carolina. (Estimated 1984 revenues, $9.8 billion; assets $20.8 billion; access lines, 13.6 million.)
- *Bell Atlantic.* Local telephone service in Virginia, West Virginia, District of Columbia, Maryland, Delaware, Pennsylvania and New Jersey. (Estimated 1984 revenues, $8.3 billion; assets $16.3 billion; access lines, 14.2 million.)
- *Nynex.* Local telephone service in New York, Vermont, New Hampshire, Maine, Massachusetts, and Rhode Island. (Estimated 1984 revenues, $9.8 billion; assets $17.4 billion; access lines, 12.8 million.)
- *U.S. West.* Local telephone service in Oregon, Washington, Idaho, Montana, Minnesota, North Dakota, South Dakota, Iowa, Nebraska, Wyoming, Colorado, New Mexico, Arizona, and Utah. (Estimated 1984 revenues, $7.4 billion; assets $15 billion; access lines, 10.6 million.)
- *Southwestern Bell.* Local telephone service in Texas, Oklahoma, Kansas, Missouri, and Arkansas. (Estimated 1984 revenues, $7.8 billion; assets $15.5 billion; access lines, 10.3 million.)

American Telephone and Telegraph provides the bulk (approximately 93% in 1984) of long-distance transport within the United States over its interexchange network (ATTIX). These services are regulated by the FCC. The corporation is free to pursue virtually any business activity, including the provision of CPE (telephones, terminals, and PBXs) and data services, through AT&T Technologies, which will continue to provide equipment to domestic telephone companies, and will market overseas. Competition for long-distance transport services is provided by OCCs such as MCI and GTE Sprint. Subject to a minimum of regulation, OCCs will eventually have access to local exchanges equal to that enjoyed by ATTIX. At present, long-distance calling over ATTIX requires the customer to touch 1 + (area code) + 7-digit number. Over an OCC it requires (7-digit OCC access) + caller identification + (10 digit number). In addition, OCCs do not receive all call setup information, and cannot make use of CCIS facilities (see Section 4.1.2.1.1).

The 7 ROCs which have been formed from 22 BOCs provide local telecommunication services. Somewhat similar to each other in revenues, assets, and customer-access lines served, they are regulated by state utilities commissions. The companies are permitted to provide CPE and mobile telephone service and to publish directories (including Yellow Pages). A common support organization has been formed by the regional operating companies to provide technical expertise and ensure network compatibility. Called Bell Communications Research Incorporated, it is chartered to supply technical information and systems software for the planning, evolution, deployment, and operation of exchange area networks and services. Together with the local telephone companies of GTE (which serve more than 10 million customer-access lines in 31 states), United Telecom (which serves more than 3 million customer-access lines in 20 states), Continental Telecom (which serves more than 2 million customer-access lines in 37 states), and Southern New England Telephone (which serves more than 1 million customer-access lines in Connecticut), the seven regional operating companies provide basic telecommunication services to approximately 95% of U.S. customers. Figure 1.8 shows the basic telephone service areas of the seven ROCs, GTE, and Southern New England Telephone (SNET).

The separation between local and long-distance companies has resulted in changes in the cost of telecommunication. No longer can long-distance calling subsidize local service, so that long-distance rates are falling and local rates are rising to reflect the actual costs of these services. How far they will go depends on the democratic process of checks and balances. Anxious to preserve the principle of *universal* service (i.e., the local company shall serve the public at rates perceived as affordable), Congress is watchful to see that the FCC does not impose significant additional charges on residential and small business customers. The result is that the fees necessary to support local service are collected from large business customers, and the long-distance carriers. Arguing that they should not pay as high fees because they are not yet afforded equal access to

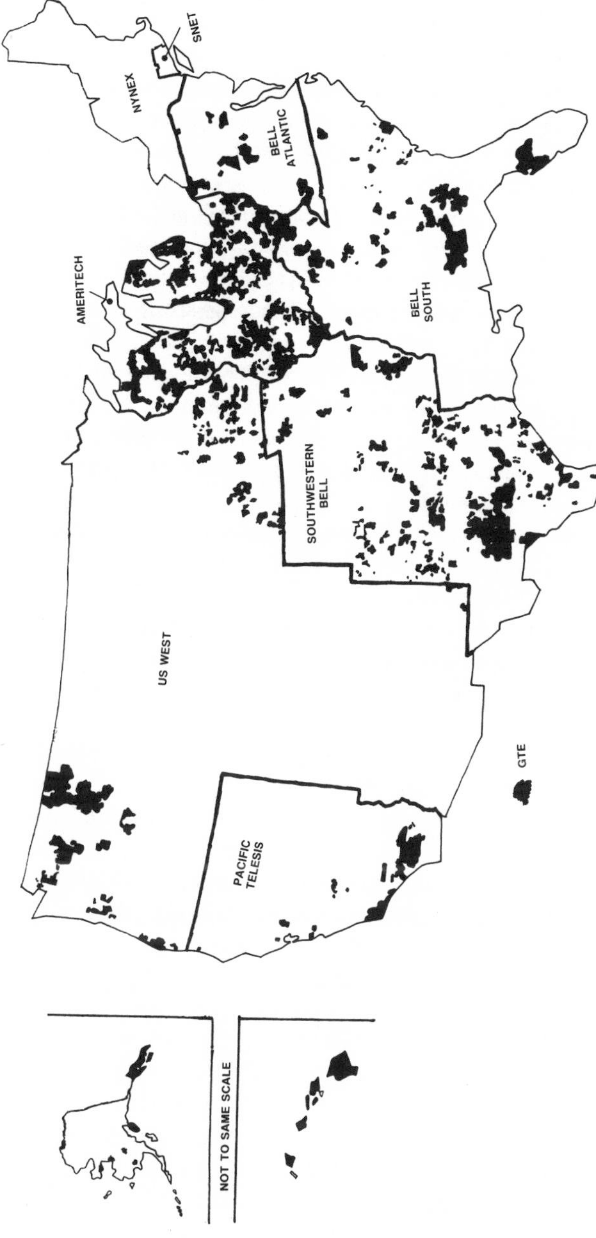

Figure 1.8 In the United States, basic telephone service is provided by some 1,400 telephone companies. The largest are the seven Regional Operating Companies (Nynex, Ameritech, US West, Pacific Telesis, Southwestern Bell, Bell South, and Bell Atlantic) and GTE. Each serves more than 10 million access lines. In addition, United Telecom serves more than 3 million customers in 20 states, and Continental Telecom serves more than 2 million customers in 37 states. Southern New England Telephone (SNET) serves over 1 million customers in Connecticut.

the exchange area, OCCs have managed to keep their payments lower than those paid by ATTIX. This permits them to compete for a larger share of the long-distance market with lower prices. In reality, regulation of this sort will continue indefinitely: a totally free market in local and long-distance telecommunication is unlikely.

What is certain is the proliferation of service and equipment providers that will compete for traffic at all customer levels. The size of the market, the rapid development of new technology, and the freedom to compete will ensure that many ways will be offered for the cost-effective transport of voice, data, and video messages to the office and the household. Price and performance will assume increasing importance as customers become aware of the options available to them. Companies who seek to provide alternative transport services will first concentrate on the largest business users seeking to persuade them to *bypass* the local serving companies. Using facilities that connect directly to long-distance networks, they will be able to offer money-saving opportunities to those who must communicate heavily among several locations around the country. In the largest businesses, there is a high potential for bypass; some 20% have already availed themselves of this opportunity, and another 30% say they plan to do so. It is a serious threat to the established operating companies, which they will combat vigorously, because only 1% of all customers account for 20% to 30% of their entire revenues.

In the home, alternative transport may be offered by cable television, and new telecommunication modes will be required for personal computers and information retrieval services. Telephony, television, and data communications will exist alone or in combinations in facilities which will be distinguished by their variety. Terminals from one, local service from another, and long-distance transport from a third supplier will be a commonplace—and they will be changed out as applications evolve and new technology reduces costs or improves performance. With no central source to select, obtain, install, and test such systems, the customer must become the system integrator. This book is directed to that subset of business and residence customers who wish to know more about current developments and possible futures so that they can understand the new environment and the telecommunication opportunities available to them.

2

Media

Over many lifetimes, building on the concepts of languages, alphabets, and numbers, media have been developed to permit the transfer of information for professional, commercial, industrial, and personal purposes. In this process, they have increased capabilities, affected social values, and modified society. Because the development of media is one of evolution and replacement, this chapter begins with a concise overview of existing record and real-time media. It then describes the new telecommunication media which provide opportunities to retrieve information, exchange messages, and conference in various ways, and discusses their applications and implementations. A final section addresses the question of other telecommunication media.

2.1. The Message of Existing Media

The characteristics of selected media are shown in Figure 2.1. The change of scale, or pace, or pattern they have introduced into human affairs has been called the *message* of the medium.[1] In presenting this summary of concepts which are qualitative, not quantitative, I have elected to express my own opinions in the expectation that they may stimulate the readers to form their own. My purpose is to illustrate the changes which existing media have created in our environment and to suggest that new media are likely to effect other modifications (which, as yet, are unknown). The discussion is divided into record media and real-time media.

2.1.1. Record Media

Books, photographs, newspapers, phonograph records, movie films, video and audio tapes, letters, and telegrams are examples of record media (which preserve a limited amount of information for future reference by the user). Each has affected the social environment.

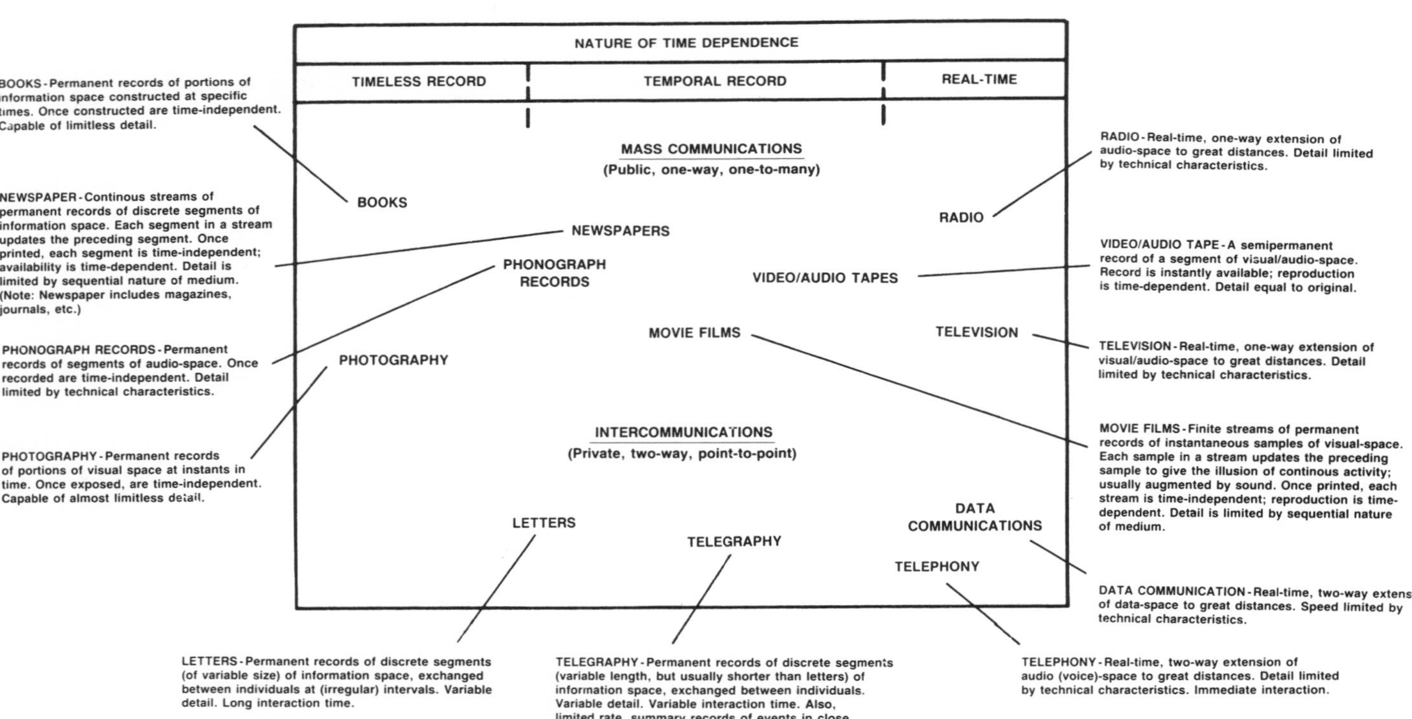

Figure 2.1. Characteristics of selected communication media.

2.1.1.1. *Books*

By making knowledge available to all, printed books stimulated literacy, and encouraged universal education. They freed learning from the feet of the medieval professor and opened to all who could read the sources of knowledge once limited to those who surrounded kings, princes, and bishops. Books represent a practical vehicle for the accumulation of the experience of one lifetime, so that it can be studied, refined, and expanded in the next lifetime. In this way, ideals are articulated, and ideas are distributed so that successive generations become more sophisticated, institutions develop, and industries grow. Books are fundamental to the growth and development of technological civilization.

2.1.1.2. *Photography*

Photographs make identification easier and allow the positive reconstruction of events. They can be a vital accompaniment to words, can substitute for written descriptions, and can add precision to the recording of technical knowledge for the use of future generations. Photographs have made lifelike paintings obsolete, and precise descriptive prose unnecessary.

2.1.1.3. *Newspapers*

An individual copy of a newsletter is an element in a stream of knowledge providing status reports on events as they unfold. Newspapers deliver a neatly packaged summary of news and views as part of the daily routine which become ready topics for conversation, and draw attention to change on a large scale. These news capsules produce anxiety over things that the individual reader cannot possibly alter, inform of local opportunities, foster a sense of community, and promote concern for honesty and fair play.

2.1.1.4. *Phonograph Records*

Phonograph records capture segments of audio space, so that they can be reproduced at will at any future time. They have made first-rate musical performances universally available, with the result that music making is no longer a common activity. Only the best is good enough, and few can compete with the Boston Symphony or the Beach Boys.

2.1.1.5. *Movie Films*

A movie film is a series of photographs taken and displayed rapidly, so as to give the illusion of motion. Combined with a sound track, they create a vehicle which frees stories from books, sometimes at the expense of detail. Movies generally cater to the fantasies of the American dream, influencing behavior and fashions, and creating glamorous stars from unknown people.

2.1.1.6. Video/Audio Tapes

Recording tapes produce records which can be replayed at will to check what was said, or done. For instance, the instant replay allows millions of football fans to check the referee's call, and a candidate for political office must be careful to be consistent, for the opposition will seek to embarrass him with recorded examples of his inconsistencies. There is no longer any uncertainty surrounding important actions: recording tape has brought a new dimension to personal accountability.

2.1.1.7. Letters

Letters, which have some of the attributes of books and newspapers, are private communications between individuals. On a personal level, they are status reports exchanged between relatives or friends which sustain a sense of social intimacy and of belonging to a separate community. In business, letters carry information which makes it possible for organizational entities to function in the larger world of the corporation. Some letters contain commercial messages and are an adaptation of a highly personal medium to mass communications. It is no wonder that these are referred to as junk mail.

2.1.1.8. Telegraphy

At one time, a telegram was the fastest way of sending a message between two points. Used for only the most important messages, and paid for according to the number of words used, they gave rise to a short, direct style of writing pseudoprose. Today, telegrams are hardly faster than the mail, and more expensive. For commercial messages, telegrams are used by establishments interconnected by a special network (telex network) in which delivery is automatically accomplished through a printing terminal. (Later developments are discussed in Section 2.3.2.1.2.)

2.1.2. Real-Time Media

The real-time, electronic, telecommunication media (radio, television, telephone, and data communication) have had a profound effect on the way in which we live. The modern household depends on them for entertainment, information, social contacts, and assistance. Radio, a broadcast medium, is available virtually everywhere, at any time, providing entertainment and information which can be listened to attentively, but which is more often used for background noise and to provide company. In contrast, television will not stay in the background. It captures the visual sense with a changing image and demands attention. That is, until the telephone rings, signaling a request for person-to-person communication over any distance—a request which is hard to refuse. In the modern office, the transaction of daily business is vitally dependent on the telephone and data communication.

2.1.2.1. Radio

A mass communication medium, radio extends audio space to great distances. Simultaneously, one person can be heard by millions. Radio emits a continuous stream of information which can be received anywhere, giving advice and instructions, or just filling in the background, relieving boredom, and providing artificial company. Radio increases the pace of news reporting; bulletins instantly involve millions of people in events as they happen. But timeliness is bought at the expense of detail. A news story must be reduced to a hundred words—or less—to fit the fast-moving format the broadcasting networks have established. As a result, we must be ready at the exact second of the hour if we are to receive our full allotment of information. If we are a few minutes late, it will have passed by, to be replaced by other messages. This preoccupation with time has spilled over to daily activity: radio has made us aware of exact time, and we organize our lives accordingly.

2.1.2.2. Television

Like radio, television is a mass communication medium which extends audiovisual space to great distances. Simultaneously, an individual can be seen, as well as heard, by millions. Unlike radio, television will not stay in the background; it presents a changing image and occupies the visual sense by producing neuronal activity in a significant fraction of the brain. It is a fascinating medium for the young, and has done much to impart general knowledge to preschool-, and school-age children. It is also a fascinating medium for grown-ups. Without effort on their part, it provides enough audiovisual information within arm's reach to keep them entertained. But it also keeps them at home, and changes the nature of their social activity. Social intercourse is diminished, a sense of immediate community is dissipated, and even the level of family interaction is decreased as each becomes preoccupied with the television set. It also creates temporary communities of millions who are united by watching history made (the funeral of President John Kennedy, or NASA's moon flights, for instance, or the World Series, the Superbowl, or the Olympics). Like the newspaper, television brings news of world events to viewers, and reinforces the sense of exact time created by radio. More than any other medium, television makes viewers aware of deficiencies in their lifestyles. To have elegance and comfort so intimately displayed in their own homes, creates an awareness of the shortcomings of their existences. Insofar as the television screen constantly displays bizzare situations designed to capture and retain viewer interest, it stimulates the modification of traditional values and activities.

2.1.2.3. Telephony

By coupling directly to the human voice, the telephone offers an easy-to-use, natural extension of private audiospace for person-to-person communication over any distance. By providing information on demand, and making assistance available in time of trouble, telephones help combat the growing segmentation of urban society. By forming flexible links between people and their increasingly mobile world, the telephone provides them with an opportunity to maintain the intimate contacts necessary for good mental health. By making everyone as close as the handset, the telephone makes it unnecessary for them to talk face-to-face with their neighbors. Their mental community can exist from coast to coast, and around the world. Because decisions can be appealed to higher authority with a simple telephone call, telephones have contributed to more equitable treatment for people: they have also made it possible for a caller to be passed from one office to another without gaining satisfaction and for calls to be ignored. Without telephones, the modern corporation with dependent operations in many cities (perhaps around the world) would find it impossible to continue to operate. The same is true of modern government, public institutions, social organizations, and much of the infrastructure of modern life.

2.1.2.4. Data Communication

Data communication transports messages to accomplish machine-to-machine telecommunication. It offers an extension of private data space over any distance, making big government, big institutions, and big corporations possible through the sharing of information. By facilitating the flow of records within and between organizations, data communication contributes in a major way to the survival of the sophisticated economies of developed countries.

2.2. Evolution and Replacement

Communication media augment the human activities of speaking, drawing, counting, and writing as shown in Figure 2.2. They have evolved from one another with the onward march of civilization and the development of new technologies. Each new medium has taken something from the existing media, and has often replaced the older media. Thus, the human faculty for drawing

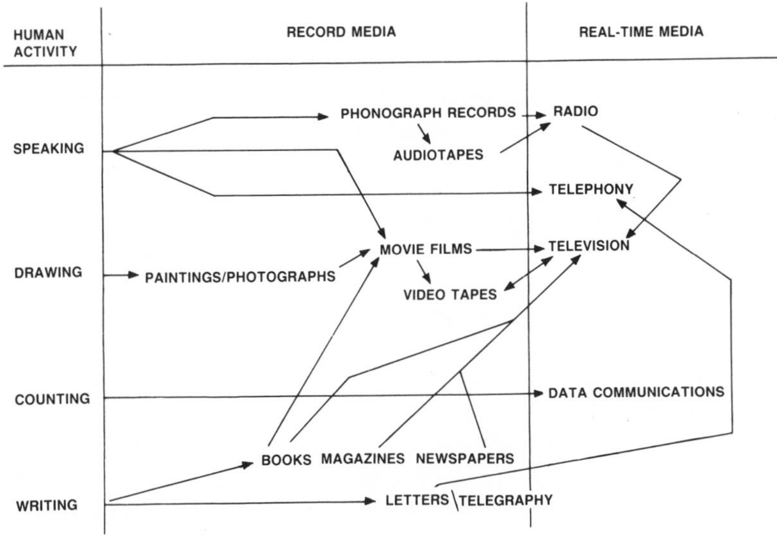

Figure 2.2. Evolution and replacement of media. Human needs and technical opportunities have created media which assume some, or all, of the function and content of other media.

resulted in the development of painting skills represented by the prehistoric art of Pyrenees cavemen or the exquisite detail of the Renaissance. However, with the introduction of lithography and, later, of photography, lifelike paintings were no longer the only way to record the sight of important people and events. The artist was left free to soar over the world of the eye to explore the dimensions of the mind. Currier and Ives were not as fortunate: their painstaking work has simply been replaced by Kodak and Polaroid.

The development of writing led to manuscripts—and printed books. Further technical improvements made magazines and newspapers possible, providing periodic installments of human drama which had the appeal of timeliness and the suspense of waiting for the next edition. But the written word is a poor substitute for the real world. By combining sequential photography with sound, movie films brought stories to life, giving the semblance of space and time, and replacing the reading of the printed page by visual experience. In their turn, movies have been displaced by television, and, in the same way that books became the contents of many movie films, films have become the contents of much of television. The same has happened to phonograph records, and their companions, audiotapes—they captured the audible arts, and have become the predominant content of radio. In its turn, radio has been partially displaced by television.

Turning to personal communications, letters have been used for thousands of years. From the clay tablets of the ancient Egyptians, we have progressed through the bark scratching of New World Indians, and the quill pen of our recent forebears, to the ball-point and printed notepaper of today. In the middle of the 19th century, priority personal messages began to travel over copper wires—the telegraph was born—and later in that century, people were able to talk to each other over great distances on the telephone. This development revolutionized personal communication, serving to extend a person's voice and the ability to hear the ends of the earth so that letter writing is now a dying art, having been replaced by the intimacy of the telephone call. And the telephone has done more, for the telephone network links computers and data processing devices into networks, allowing machine to communicate with machine and perform many functions automatically under the control of programs prepared by human analysts. Data communication is having as much effect on the details of commerce as the telephone has had on human communication.

Until the creation of the real-time communications media, the information delivered was in the past tense. For instance, books and photographs preserve information for all time. Once created, their contents are history, and they exist independent of the passage of time. Newspapers, phonograph records, video/audio tapes, movie films, letters, and telegrams also contain historical information. As media, however, their effect depends on time in one way; time is involved as one segment succeeds another. In contrast, the telecommunication media can deliver information anywhere the instant it is produced. The price of this spontaneity is that the information is not preserved for reference. Nevertheless, in most of our lives, the telecommunication media have created a greater dependence on them than any of the other communications media. They are the foundation on which the media of the future will be built.

2.3. New Telecommunication Media

New media are being developed which produce added dimensions to the telecommunication process. Together with today's real-time communication media they are expected to facilitate the application of electronic technology to improve productivity in commercial and industrial organizations and to encourage a broad spectrum of services in households of the future. The relationship between today's real-time media, existing record media, and the new media is illustrated in Figure 2.3. The new record media are: *videotex,* which is implemented in two ways, as wired videotex and broadcast videotex; and *electronic mail,* which exists in text, voice, and graphics forms. The new real-time media are: *teleconferencing,* which exists in computer, voice, augmented voice, and video forms; and *interactive television,* which provides many video channels to the customer, and a limited return capability.

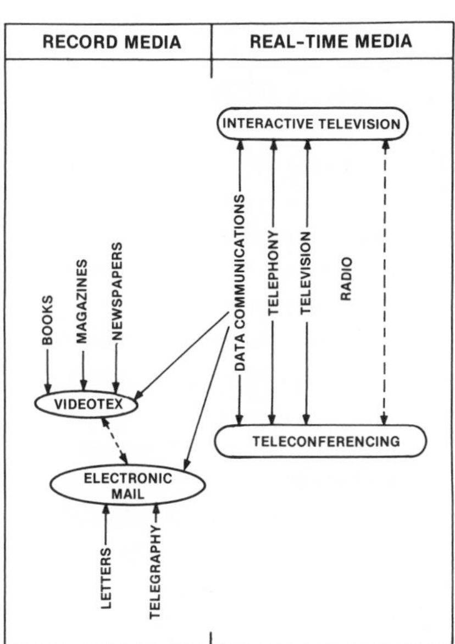

Figure 2.3. Relationship between new and existing media. Books, magazines, newspapers, letters, and telegraphy are existing media which record capsules of information of interest over different time scales. Data communication, telephony, television, and radio are real-time media which deliver information the instant it is received. Videotex, electronic mail, teleconferencing, and interactive television are new media which derive some of their functions from existing media, and also from themselves. © 1981 IEEE.

2.3.1. Videotex

Videotex is a medium which facilitates information retrieval by many persons from remote databases for business or residential purposes. Videotex is implemented in two ways—as wired videotex, which is a two-way, fully interactive medium, and broadcast videotex, which is a one-way medium with selection.

2.3.1.1. Wired Videotex (Viewdata)

Wired videotex is a two-way, interactive system which uses a video display (often a television receiver), local processing, and a remote database accessible through the public telephone network. It is often referred to as *viewdata,* the name of the pioneering development by British Telecom (in the early 1970s): often it is simply referred to as videotex. As a medium, it delivers pages of information, each of which fills the video screen. They are stored in a remote datasbase ready to be accessed by employing a search protocol which allows the user to page to increasing levels of detail. The database can embrace a large number of subjects; for this reason the medium has been likened to an electronic encyclopedia. Its contents are derived from books and newspapers, and it includes some of the functions of data communication.

Wired videotex is supported by existing telecommunication suppliers who foresee applications to both business and residence sectors. Based on the use of in-place equipment supplemented by additional units which incorporate advanced technology, viewdata can provide information from many unique databases of interest to business alone (e.g., parts catalogs, credit information), some databases which will have application to both business and residence sectors (e.g., Yellow Pages, timetables), and some databases which will be of interest to residences alone (e.g., shopping catalogs, school information). Figure 2.4 shows the concept of a typical system and a list of some of the subjects contained in current databases.

2.3.1.2. Broadcast Videotex (Teletext)

Broadcast videotex is a one-way information system which employs an unused portion of the broadcast television signal to transport data to modified television receivers without interfering with the normal program. It is often referred to as teletext, the name of the pioneering development by British Broadcasting Corporation (in the early 1970s). A sequence of pages of information is sent repetitively. From them the user can select pages of interest. On average, he must wait for several seconds until the page he seeks is transmitted and his receiver has captured and displayed it. To make this time acceptable, the number of pages in the sequence is limited to no more than a few hundred. Thus, the detail of wired videotex is not possible, and the service is often described as an electronic magazine. Its contents are derived from magazines and newspapers, and much of what teletext contains may be included in viewdata.

Using in-place equipment supplemented by additional units that incorporate advanced technology, broadcast videotex is supported by existing telecommunications suppliers. In principle, teletext can deliver information of interest to both residence and business sectors. In practice, it is likely that the limited amount of information available will make business users resort to wired videotex or other media. Figure 2.5 shows the concept of a typical system and a list of some of the subjects contained in current databases.

2.3.1.3. Applications

2.3.1.3.1. Viewdata and Teletext. Both viewdata and teletext retrieve information from the remote database by a tree search routine. In this technique a menu is displayed which lists the main categories of information available and identifies each with a number which the inquirer must press on the keypad to continue the search. Pressing the appropriate key causes a second page of information to be displayed. It gives greater detail on the subject selected from the menu and identifies the numbers which must be used to call up the next pages, which contain finer detail yet. Using this routine, it is possible to reach whichever part of the tree contains the information of interest. The search can

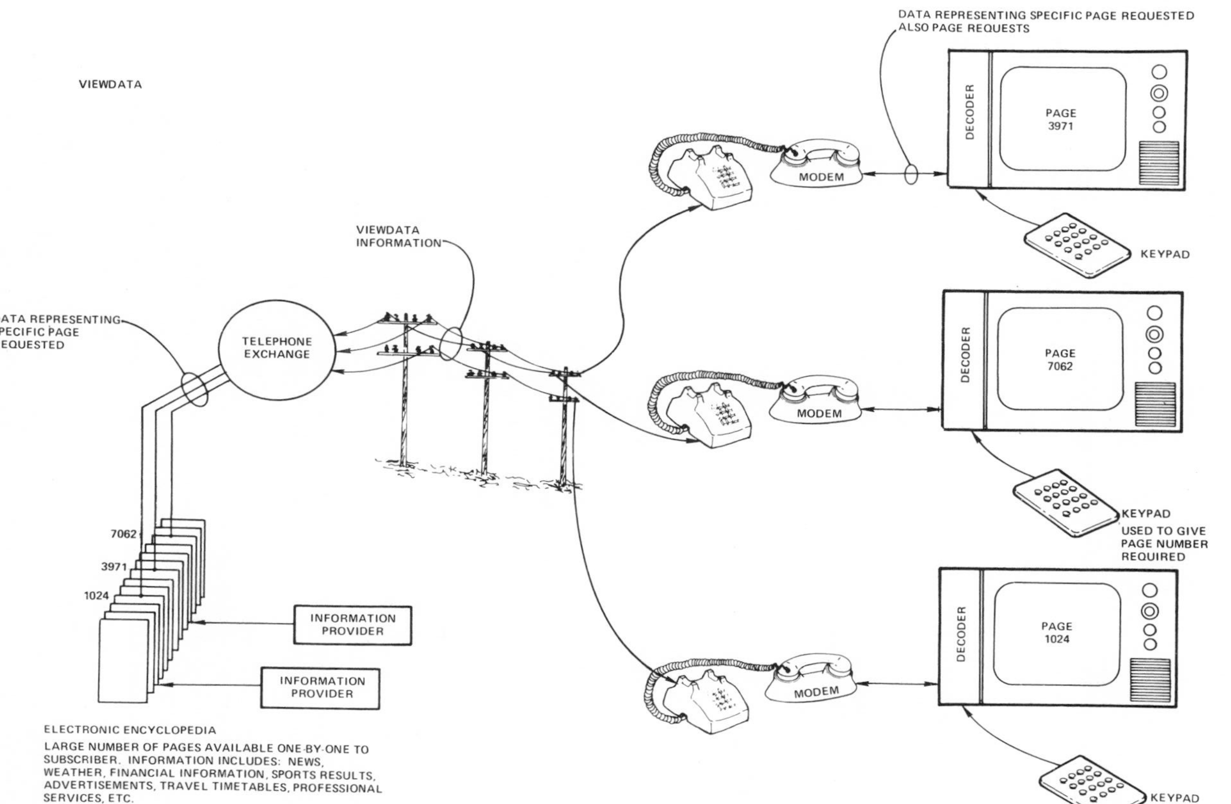

Figure 2.4. Architecture of a typical wired videotex (viewdata) system. The user at home, or at the office, is connected through the telephone network to databases in the videotex network. With a fully developed system the user could conceivably access information from any center, worldwide.

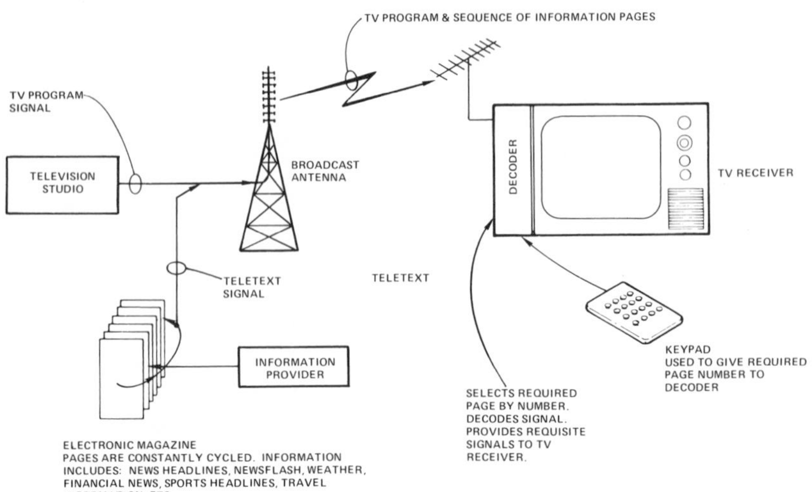

Figure 2.5. In a broadcast videotex (teletext) system, pages of information are arranged in sequence. They are added to the standard television signal. The user identifies the page he wishes to consult, then waits for the receiver to capture and display it. © 1981 IEEE.

be shortened by requesting a specific page by number, and there is usually an escape command which will return the inquirer to a given level of the tree should the search go in the wrong direction or become too tedious. Special items of interest may be highlighted at the beginning and page numbers supplied so that the inquirer can retrieve them directly.

2.3.1.3.2. Viewdata Only. Because it is a two-way medium which connects to a computer-controlled database through the telephone network, viewdata can support extensive interactive services. For instance, shopping from home is possible—for merchandise, whose specification is called up, general features are displayed, availability indicated, price listed, order accepted, and charged to a credit card, or paid for by funds transfer—or for transportation, whose timetable is called up, a specific run selected, space availability indicated, reservation made and confirmed, and paid for. Other possibilities include the acquisition of large blocks of information for later use—an electronic library, or the acquisition of computer programs—for business, educational, and entertainment purposes. In its simplest form, wired videotex has been designed to operate with a 10-button keypad. Provision of a full alphanumeric keyboard makes more sophisticated searching and interaction possible, including sending free-form messages from terminal to terminal.

2.3.1.3.3. Teletext Only. Besides providing information services of general interest, broadcast videotex can be used for special tasks in which information added to the normal signal is only of interest to limited groups of viewers.

Examples are captioning television programs for the deaf, and inserting foreign language subtitles. In these cases only those wishing to do so need display the additional information.

2.3.1.4. Implementation

In 1983, wired and broadcast videotex systems were available as public services—or were in field trials—in about 20 countries. More than one million terminals existed: one-third of them were to be found in the United Kingdom and one-quarter of them were in France. A few were in use in trials in the United States.

2.3.1.4.1. Viewdata. In the 1970s, telecommunication organizations in the United Kingdom, France, Japan, and Canada developed wired videotex systems which employ modified television receivers and incorporate unique data presentation and coding features. In the United Kingdom, viewdata is offered to the public under the service name of *Prestel.* Pages of information can be acquired every eight seconds, on average. In France, viewdata is exploited under the general name of *Titan.* While different in detail, the result is generally similar to Prestel. In Japan, viewdata was developed under the designation *Captain.* Because written Japanese consists of some 3000 Kanji, Katakana, and Hiragana ideographs, not a limited, sequential alphabet, the display is produced in facsimile fashion (as a set of individual dots). In Canada, viewdata was developed under the designation *Telidon.* It employs a page-coding technique which produces a combination of text and pictures whose resolution is limited only by the resolution of the terminal used.

Inevitably, the quality of the graphics produced by the early systems developed in the United Kingdom, France, and Japan suffers in comparison with the more sophisticated presentation of the Canadian system (Telidon). It showed that flexible combinations of text and pictures could be prepared which are pleasing to the user, and more useful to the service provider. However, even Telidon is not perfect, and in 1981, AT&T introduced the Bell System Videotex Standard Presentation Level Protocol. The Presentation Level Protocol (PLP) incorporated all of the features of the earlier systems to enable existing databases to be used (with the aid of electronic manipulation), and included a broader range of features to make the end product acceptable to potential sponsors. These additions permit a wider range of colors and the inclusion of refined graphics coding to allow the realistic reproduction of company logos (or other intricate designs) and facilitate the repetitive use of these items. *De facto,* PLP has become the North American Standard for videotex[(2)] and has stimulated the designers of the earlier European systems to upgrade their graphics capabilities. (More information on wired videotex is given in Section 4.4.3.)

2.3.1.4.2. Teletext. Telecommunication organizations in the United Kingdom, France, and Canada have developed broadcast-videotex systems. In North

America, early trials used systems based on British and French techniques. More recent activities are centered on a PLP-compatible format. Each employs the same data presentation and coding standards as are used by the national viewdata systems. With any of these systems, the user must wait 10 to 15 seconds (on average) to view the page selected from a total of 100 pages. (More information is given on broadcast videotex in Section 4.4.3.)

2.3.1.5. *Needs and Motivations*

Videotex is readily available in only a few countries. Presently, the high cost of electronics required to convert standard television receivers to display terminals, the ready availability of more familiar reference sources, and the lack of a clear need that videotex can assuage better than anything else, are hampering the expansion of the service. For instance, when the user seeks information normally contained in the newspaper, more than a 15 to 20 second wait at the terminal—while the receiver is switched on and warms up, the connection is made, or the proper page is seized—may be enough to discourage electronic searching in favor of leafing through the newspaper. However, such a wait is acceptable if the information can only be obtained by going to the public library. While the tree search technique is easy to comprehend and implement, it can become quite frustrating if the steps are too shallow, causing many pages of information to be sorted through, or the search entails many different topics, each of which must be traced from the beginning. There are other problems, too. What if members of the household are using the television receiver for broadcast reception or video games, or the telephone is in use? It may take an urgent, or insistent, need to overcome the natural reluctance to purchase another receiver, or order another telephone channel, to have videotex available on demand.

The cost of the service will play a part in its utilization. For teletext, which is broadcast to everyone, use of the service is essentially *free*. It will be paid for by the user in the form of a license fee or by the service provider who will expect a return from increased business. Demand for the general information in teletext will be determined by its usefulness, timeliness, and the ease with which it can be obtained. For viewdata, information exists within the system to identify the using terminal so that billing can be accomplished on a page-by-page basis. Both quantity of data accessed and connect time can be taken into account. Some information may be free—although the telephone connection may not. Certainly, such things as business and professional announcements, electronic Yellow Pages, entertainment listings, and travel arrangements, are likely to be free. Access to other information, such as an electronic encyclopedia, tax advice, and stock prices may not be free, and games, message services, and similar activities will not be free. Many persons believe they would subscribe to viewdata services if the monthly fee is about the same as that charged for cable television or basic telephone service.

Once the novelty of an electronic information system has worn off, most inquiries center around amusement, recreation, and hobbies. Where public transportation is important, inquiries for schedule and service information run fairly high. In some tests, system use averaged one hour or so per week. To its supporters, videotex can make major contributions to improving the quality of life by reducing time spent shopping, allowing the conduct of personal business from home, making working at home possible for information workers, and providing knowledge and education. To existing communications providers, videotex represents an opportunity to provide a new service over mostly existing equipment. To entrepreneurs the demand has yet to be verified and the return is uncertain. At the moment, videotex is influenced much more by technology push than market pull.

2.3.2. Electronic Mail

Electronic mail is a medium which facilitates the non-real-time exchange of text, voice, and graphics messages among persons. In computer-based mail systems, messages are composed, transmitted to an intermediate location, and stored on command from the sender to be retrieved, transmitted to the receiving destination, and delivered on command from the recipient. Such *store-and-forward* systems handle text and voice. Facsimile systems handle graphics.

2.3.2.1. Text Mail

2.3.2.1.1. Electronic Mail. In an electronic mail system, the originator types his messsage on a data terminal connected through the public switched telephone network, a public packet network, or private connections, to a computer which provides support functions for password verification and editing and assembling the message, interprets the address and other instructions, and forwards the message to an electronic storage unit where it is inserted in the field assigned to the recipient. This serves as an *electronic mailbox*. The originator may send the message to a single person or several persons, request confirmation that the message has been delivered, and file a copy. When the recipient connects to the system and presents the correct password, the system displays any broadcast messages, then indicates that a message or messages are waiting. The recipient may elect to review the senders' names and the subjects, and read some, or read them all as they are presented by the system. Each message may be deleted, filed, forwarded, answered, or processed in some other way.

Figure 2.6 shows the concept of an electronic mail system. Besides simple data terminals, messages can be inserted and retrieved through local computers

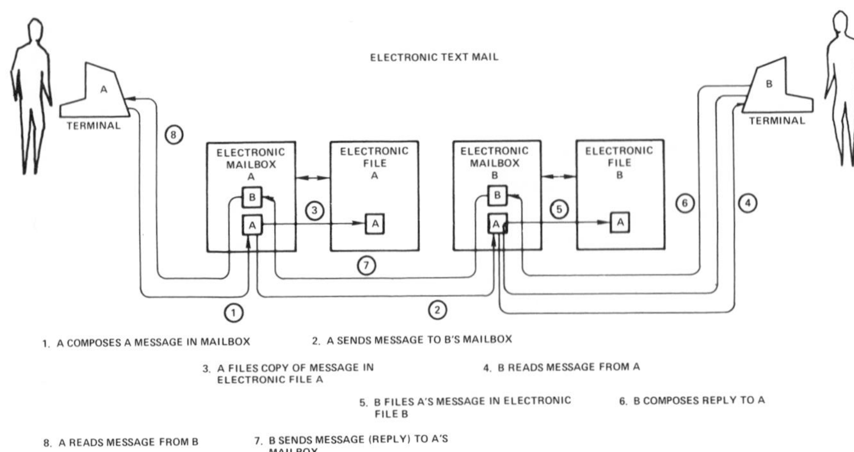

Figure 2.6. Block diagram of an electronic mail system.

and composed, edited, and formatted using indigenous word processing capability or word processing equipment. For long messages, reports, etc. preprocessing of this sort reduces the connection time to the network computer.

2.3.2.1.2. Telex/Teletex. Prior to the invention of computer-supported text message systems, a message service was provided by communicating teletypewriters and teleprinters. In the simplest form, the operator would compose the message at the teletypewriter keyboard recording it on punched paper tape. The operator then contacted the intended station and transmitted the message by passing the paper tape through the reader on the teletypewriter. If direct communication was not possible, the operator passed the message to an intermediate station where another paper tape of the message was made. The second tape was used to send the message to the destination at a later time. Often, the message was combined with several others going to the same destination. Called telex service, it began some 50 years ago and has developed into a worldwide compatible text communication service with some 1.5 million subscribers. Improved over the years by the introduction of some automatic features, it still provides slow, but reliable service. In much of the world it is more readily available than telephone service.

In 1983, recognizing a demand to link more sophisticated terminals, several telecommunication organizations inaugurated a service called *teletex* (not to be confused with telete*x*t which is discussed in Section 2.3.1.2). International bodies have agreed on terminal standards and the code to be used so as to permit general communication among users of electronic storage typewriters and communicating word processors. At a teletex terminal texts are composed, edited, and stored, and the terminal may be used for local functions independent of the transmission or reception of messages. The basic character set includes all letters, figures, and symbols used in languages employing Latin characters. The user may add national characters as desired. Text to be transferred to another terminal is assembled in memory and transmitted automatically once circuits have been established between the communicating terminals and the *send* instruction has been given. A page of text can be transmitted in around five seconds. In some respects, teletex is a distributed intelligence, text mail system—in contrast to electronic mail, which is a centralized system. In teletex, processing and storage are accomplished at the terminals which communicate one with the other. In electronic mail, the service is provided through a central processing unit.

2.3.2.1.3. Structured Message Processing. Text mail systems deliver messages to the recipient without regard to their content. The reader who examines them may be stimulated to perform an activity which results in another message, the transfer of information to a database, or some other action which involves data. By structuring the message in a manner known to the system, the originator can provide sufficient information for the receiving node to operate on the message without human intervention. In this way, messages may be forwarded to other locations on the basis of content, may be collected together and action taken only when a related set of messages is assembled, or used to generate other messages on the basis of the current state of a database. The result is the integration of message and database facilities to achieve an increasing level of office automation.

2.3.2.2. Voice Mail

Voice mail service records the spoken words of the sender, stores his message, and delivers the sound of his voice to the recipient when he calls. The service may also provide general announcements, be used for message collection, furnish call answering capability both for no-answer and line-busy conditions, automatically deliver messages and keep trying until the message is accepted or refused, and be used for recording both sides of a regular telephone conversation. Control is exercised through the touch-tone buttons on the telephone. The commands include such functions as start/stop/pause, back-up/move ahead, playback, erase, record, and send. The system prompts both sender and recipient, asking questions of the user to elicit essential information which the user keys in through the touch-tone buttons of the telephone. Figure 2.7 shows the concept of a voice mail system.

2.3.2.3. Graphics Mail

For more than 100 years, sending and receiving pictures over wires has been accomplished by devices known as *facsimile* machines. Through the years, the quality of reproduction and the speed with which a copy is made have steadily improved. In recent years, the desk-top facsimile transceiver has become an important adjunct in business. Today, equipment which can process, transmit,

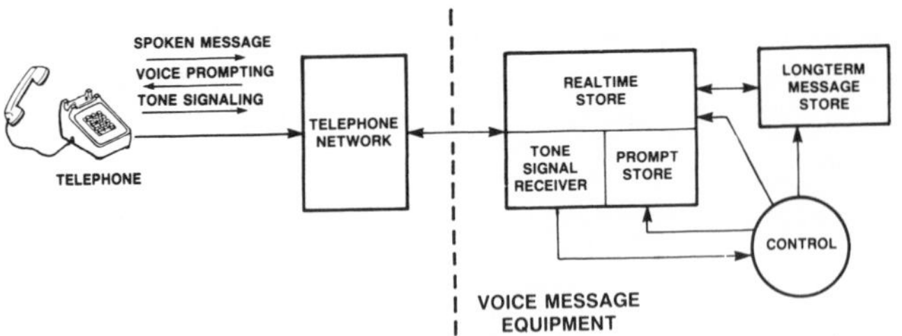

Figure 2.7. Block diagram of electronic voice mail system.

receive, and reproduce standard size letters and drawings in close to one minute is common, and higher-speed equipment is available. They are particularly useful in countries where the written language employs ideographs. In Japan, for instance, facsimile communication is much more important than electronic text mail communication.[3]

While facsimile communication has been mostly on the basis of a direct connection between compatible machines through a switched network, all of the characteristics of store-and-forward which have been enumerated for text and voice mail systems can be employed. Of these, delayed delivery, multiaddress delivery, and mailbox delivery may be of most interest to users. In addition, it is possible to provide conversion between the code used by one machine and the code used by another to allow graphics messages to be transferred between incompatible machines. (Facsimile terminals are discussed further in Section 4.6.4.)

2.3.2.4. Applications

Electronic mail can serve most of the non-real-time message needs of businesses and households. It has the advantage over the telephone that no one needs to be there to answer, and over normal mail that delivery is made at electronic speed. Why then is it not universally used? Perhaps the main reason is cost—cost of terminal equipment, cost of message storage and processing, and cost of transmission—when compared to the cost of regular mail. Also, some equipment is complicated to use, the service is not as reliable as it should be, and, in most cases, the quality of the output is not as high as it could be. Incorporating additional capability to correct these deficiencies increases the cost of the terminals as well as of storage and processing. The result is that electronic mail is chiefly employed in organizations where terminals are also used for other things, in organizations where the speed with which transactions are accomplished is important because it saves money, and in organizations where an unanswered telephone call may mean loss of business. Thus applications are chiefly found in high-technology, computer-based industry, in electronics and computer science departments of universities and research laboratories, in large financial institutions, and in professional and consulting offices. The penetration of electronic mail into the household is minor—and is likely to stay that way until terminal costs decrease drastically, an effective service is built on videotex, or postal charges increase significantly.

2.3.2.5. *Implementation*

Text mail services are offered by a number of North American organizations which route messages over telephone and data networks to central complexes where they are processed and stored in electronic mailboxes until the recipient requests they be forwarded or they are delivered under other arrangements. These services have been extended to other areas of the world. In addition, in many countries, telex service is being upgraded to teletex service.

Voice mail services are offered by a number of North American organizations which route messages over telephone lines to central facilities where they are processed and stored in electronic mailboxes to await a request for delivery or they are handled under other arrangements. In addition, expanded custom calling services in advanced electronic switches include voice mail features.

In 1982, the United States Postal Service (USPS) began to offer an electronic mail service (E-Com). Initial service is directed to customers with high-volume mailings (200 copies minimum). From a data terminal on their premises they enter messages and addresses over existing telecommunication facilities to one of 25 USPS computers. The USPS prints the messages, inserts them into envelopes, and delivers them through existing first-class mail facilities in the normal way.

Until recently, graphics mail service has been a matter for individual companies and organizations. A large number of desk-top facsimile units are scattered throughout industry. Connections are made on an ad hoc basis over telephone lines and other circuits as the need arises. International facsimile connections are available between the United States, Canada and the United Kingdom, and other parts of the world.

2.3.2.6. *Needs and Motivations*

Electronic mail appears to be readily accepted by scientific, professional, and commercial organizations. Usually, little difficulty is encountered in comprehending and employing the available systems. Most problems stem from computer failures and poor maintenance on facsimile and printing equipment. Many users feel that these systems provide more effective communication with peers and colleagues, particularly with those in remote locations.[(4)] The store-and-forward feature is effective when communication spans several time zones, or when colleagues work different hours. Satisfaction is gained from knowing the message will be delivered at a time when the recipient is available. Electronic mail reduces the frustration associated with telephone calls which find the recipient's line busy or the recipient away from the desk.

Text mail may be read at leisure from a video screen or printed copy. Time can be taken in composing messages, and it is relatively easy to edit them before sending. Long messages do not impose undue hardship on the recipient since specific text and the surrounding context can be studied readily and at leisure. On the other hand, voice mail demands the recipient's attention. It delivers a message in the sender's voice with all the emphasis and color he may wish to provide. It must be listened to. Hard copy is not available and detailed consideration of specific statements and the surrounding context is not easy to do. Long messages impose hardships on both sender and recipient. Thus, text mail may be more suited to lengthy arguments, position statements, and technical matters, and voice mail may be more suited to instructions, short administrative messages, and the like. For those skilled in dictating, voice mail could be an easier medium to use than text mail, although either medium can be used with the aid of a secretary who transcribes the dictated message into the data terminal, or transcribes the voice message received from the mail system on to paper for study and reflection. Voice mail has the advantage that it requires no special terminal, just a telephone, so that it can be sent or received from virtually anywhere.

Electronic mail has been adopted by certain types of businesses. Commercial offerings have been made and several services are available. It seems to fill a genuine business need for faster, more flexible communication. The future of electronic mail will be shaped by market pull, not technology push.

2.3.3. Teleconferencing

Teleconferencing can be defined as the medium which supports electronic communication between three or more persons at two or more locations. Ideally, the information space shared by persons in a conference room is extended to a distance so that they do not have to come together to achieve the objectives of the meeting. Meetings vary in purpose, size, and format from small, intimate meetings in which neither papers are exchanged, nor graphics are used, and the outcome depends solely on the faith and trust each person has in the other participants, to large forums, in which papers are exchanged, graphics are used, and views and ideas are argued back and forth by persons whose ambitions and objectives may be at odds with each other.

The effect of limiting the dimensions of the exchange by introducing electronic constraints is far from understood. It is known that the richness of the medium employed, the degree of familiarity of the participants with each other, the complexity of the topics discussed, and the number of persons and/or locations involved, all affect the level of satisfaction with the experience expressed by participants. For routine discussions between peers who are well known to each other, good quality audio channels may suffice for the exchange. For unique discussions between several levels of management in several locations, full color video may be insufficient. For complex discussions between technical persons, computer conferencing may be ideal. The spectrum of teleconferencing alternatives is shown in Figure 2.8. It extends from computer conferencing to audio conferencing, and through augmented audio conferencing to videoconferencing. The expense of computer conferencing can be *many* times audio conferencing, and videoconferencing is *very* many times as expensive as audio conferencing. Face-to-face meetings without electronic mediation represent the *richest* communication medium. Their expense depends on the distance the participants must travel, and the meeting duration.

2.3.3.1. Computer Conferencing

In a *computer conference*, each participant has a terminal which is connected directly, or over telecommunications lines, to a central computer. The computer passes messages among the participants who can address the entire group, a subset, or a single individual. The product of the conference activity is stored and available to all participants for reference as the work proceeds. In addition, there may be other data files which provide supporting information, or are the starting point for the conference. To some extent, this is electronic mail. In fact, it is electronic messaging among a restricted group of individuals for a common purpose. The computer attends to the exchange of messages, the maintenance of files, and the assembly of the body of data which is to be the outcome of the conference. Like electronic mail, the individual participants do not have to be

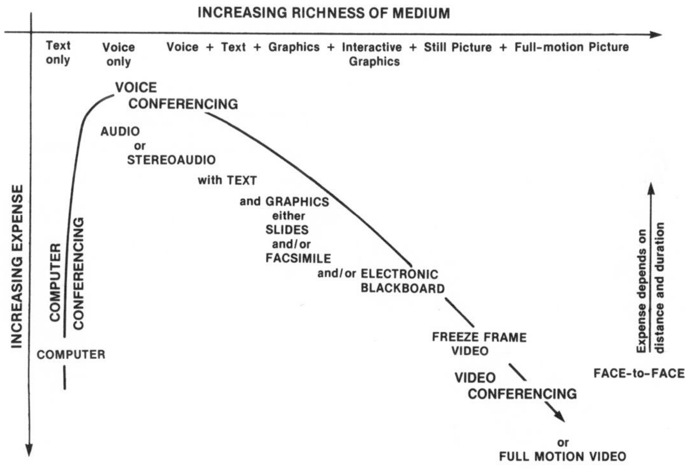

Figure 2.8. The spectrum of teleconferencing.

involved simultaneously. Unlike electronic mail, they share a message community without having to define it each time, and can operate on common information without having to build it each time. Figure 2.9 shows the concept of computer conferencing.

2.3.3.2. *Audio Conferencing*

Audio conferencing is the simplest style of teleconferencing. In relatively primitive form it has been around a long time. For groups small enough to gather around a desk, it can be implemented between two (or more) locations using speakerphones or special conferencing telephones. For those served by modern telephone switches, a three-way conference can be established unaided, and the conference can be extended one person at a time to encompass larger groups. Conferences of around six persons can be set up expeditiously by the call initiator. Connections to a larger group of persons can be made by an operator through the public switched telephone network by calling each party in turn. At a predetermined time, very large groups of persons can call a *meet-me* bridge for connection in a conference arrangement. Up to 100 participants can be connected in this fashion. Individual recognition and discussion is impossible in these large group connections without a strong chairman and a prearranged protocol.

2.3.3.3. *Augmented Audio Conferencing*

Voice-only conferences are not adequate for many purposes. When equipment or functions need to be described or the relationship among variables needs

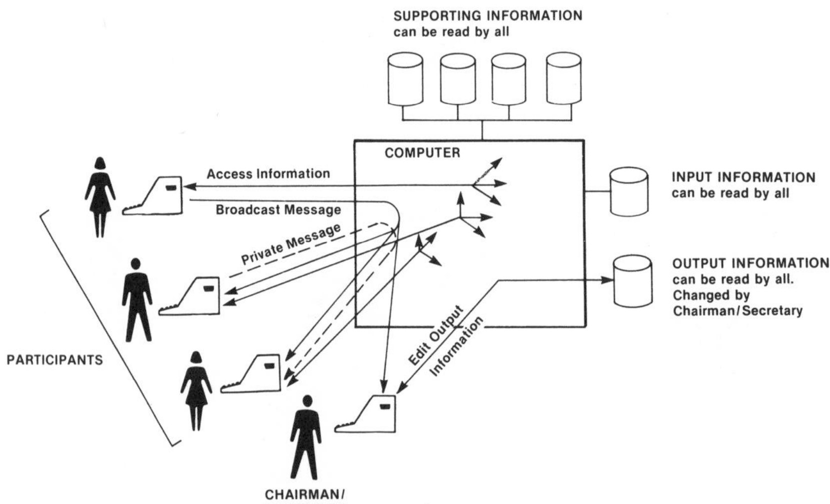

Figure 2.9. Principle of computer conferencing.

to be illustrated, there is no substitute for a picture—and when financial statements need to be analyzed, there is no substitute for properly prepared tables. These dimensions can be added to audio conferencing by employing additional devices such as slides, facsimile, electronic blackboard, and freeze-frame video.

2.3.3.3.1. Slides and Printed Materials. If there is time for preparation and distribution, slides which can be projected simultaneously in each conference location can augment the voice channel and provide equipment photographs, line drawings, graphs, and charts to which the conferees can speak. So can printed materials, typewritten exhibits, etc., although they have the disadvantage that they must be reproduced for each participant, and the meeting tends to dissolve as each person studies his own set of charts. For meetings which can be prepared well in advance and which will not need spontaneous graphics support, such arrangements could be adequate.

2.3.3.3.2. Facsimile. *Facsimile* machines can be used to augment audio conferencing facilities for meetings which require spontaneous graphics, or charts and exhibits prepared too late for mailing. The time delay in reproducing the facsimile and copying it for all to have, may be a disincentive to the free flow of ideas.

2.3.3.3.3. Electronic Blackboard. An *electronic blackboard* provides a one-way information space in which participants may write or draw their contributions. Items drawn on the transmitting board in one location will be reproduced on video displays in the other locations.

2.3.3.3.4. Freeze-Frame Video. What if the conferees wish to have periodic pictures of conferencing activities or visual aids? One technique extracts a single *frame* from a television camera operating at regular speed and transmits it over the telephone lines at a slow rate to other locations where it produces a still picture. Freeze-frame video can be used with prepared graphics, spontaneous illustrations and product exhibits, or just to give periodic views of the participants. In some versions an electronic pointer is provided which can be continuously positioned on the picture so that the sender can indicate features of the image to emphasize his presentation. Using speeds of a few thousand bits per second, freeze frame equipment can send full color pictures in a minute or less. The pictures will be of no better quality than television. Thus, they suffer from lack of adequate resolution to display any but the simplest graphics. In addition, because the picture transmitted is one selected from a full motion stream, the participants may be frozen in unbecoming poses. Fortunately, all systems give the originator control of what is transmitted.

2.3.3.4. Videoconferencing

In *videoconferencing*, extensions of audio-, video-, and dataspace connect remote conference rooms so that the sounds, sights, and motions of the occupants are available for all to see and hear. For two-location conferences such is the case. For multilocation conferences, providing full video coverage of all rooms at all times may not be practical because, unlike audio signals which can be mixed together without losing individual speaker identity, video signals must be kept separate. For n locations, then, there must be $n(n - 1)$ video connections and $(n - 1)$ monitors in each conference room to display the other rooms. Equipment and facilities become expensive in a hurry! As an alternative, a video switch can be employed so that only the scene from the conference room in which the speaker of the moment is located is distributed to all other locations. This reduces the number of video connections to each location to one two-way connection at the expense of limiting the videospace of the participants who are required to look at the speaker. For conferences among a large number of locations, two-way video may not be particularly useful.

In early systems, the picture was in black and white, and some systems still employ this mode. An important consideration is that black-and-white cameras may cost one-quarter (or less) what color cameras cost. However, most users are conditioned by their home environment to expect television pictures to be in color, and equipment displays and graphics benefit from its use. Accordingly, most modern installation feature television-quality color pictures. Using television standards has the advantage that equipment needed to implement conferencing *studios* already exists, and the disadvantage that the resolution provided is marginal for most graphics. For this reason, videoconferencing systems are usually augmented by high-speed facsimile machines to provide acceptable exchange of typewritten and similar material.

2.3.3.5. Applications

Teleconferencing can be applied to all of those business situations for which conferences are appropriate. Thus, project progress reviews, problem-solving sessions, product introductions, press conferences, etc., can all be accommodated within the arrangements described. The size and style of the meeting are determined by its objective, the characteristics of the organizer and participants, and the facilities available.

In commercial and institutional settings, teleconferencing can be used in educational and training applications. In many real-life situations the flow of information is far from balanced. For these applications it is unnecessary for the communications capabilities to be equal in both directions. University teaching is a case in point. Video and voice from the professor to all locations, with voice return, can be quite adequate.

In professional activities directed to producing a report or position paper, computer conferencing is ideal. It can be augmented with facsimile for graphics and calculations if required. Participants can compose segments of the report, read and criticize each other's contributions, have access to resources at their home locations, and ponder issues without distraction. In fact, computer conferencing may be an ideal medium for developing positions without being swayed by zealous proponents of one side or the other.

2.3.3.6. *Implementation*

Teleconferencing facilities are rapidly being accepted as components of the contemporary office environment. All modern electronic PBXs can be obtained with voice conferencing features, audio conference service is supported by electronic end-office switches, and private audio conference services are available. Most facsimile machines, electronic blackboards, and other equipment used to augment the voice channel are available with integral telephone connections. In addition, several companies have developed stereo voice channel arrangements in conjunction with specialized worktables for the participants. In short, satisfactory audio conferencing facilities can be obtained by all who wish to use them.

Computer conferences can be arranged by all who have a computer available, an operating system which supports telecommunication, and one of several commercial software packages which implement conferencing.

Videoconferencing is available to all who have the money. Casual users may schedule full-motion videoconferences using Picturephone Meeting Service (PMS) studios placed in large cities by AT&T (see Section 4.3.3). Those with a continuous demand may install studios on their own premises which can be connected to PMS, or served by other interexchange carriers. Several hotel chains have installed public videoconferencing facilities so that geographically distributed meetings can be held at their facilities.

2.3.3.7. *Needs and Motivations*

In the United States, it has been estimated that some 20 million meetings occur daily. More than half are said to last less than 30 minutes, and most do not require visual aids. For many meetings, some participants travel from somewhere else, depriving their home location of their presence, incurring unavoidable lost time, and sometimes placing unnecessary stress on the individual. Insofar as teleconferencing provides an extended meeting space, it reduces the need for travel, saves time and money, and can contribute to the overall well-being of the work force. Further, in psychology departments, extensive testing of the speed at which problems are solved has established that there is no significant advantage to face-to-face communication over communication by voice alone.[(5)]

Why then is teleconferencing so little used?

One reason may be that psychology tests are usually executed in a cooperative spirit with the participants seeking to overcome the limits placed on them. They have no alternative means available to accomplish the end. In business, cooperation is not necessarily present, and the alternative to voice conferencing is to travel to attend a face-to-face meeting—a mode which is familiar, well understood, and has worked in the past.

2.3.3.7.1. Imbalance of Sensory Channels. Another explanation may be that audio conferencing is an activity in which the participant talks to unseen persons while viewing his present surroundings. The field serving a sensitive sensory input channel (the ear) and the field serving the sensory channel which performs the most complex processing and demands attention (the eye) are not the same. For more than a person-to-person conversation, the participants find visual cues helpful in separating the speakers and may become confused when none are available in the immediate environment. Calling persons together in a conference room matches sound and sight, and restores the coincidence of the inputs. For this reason, full-motion video may be an essential component in a broad-based teleconferencing system intended to be used in all situations.

2.3.3.7.2. Richness of Medium. A medium is said to be rich to the degree that it mimics the communications capabilities manifest in a face-to-face meeting (which is the richest medium). The richness of the medium has an effect on the size of the group and the complexity of the subjects which can be handled effectively in a teleconference. With an audio-only link, a medium of little richness, more than a few persons discussing a complicated or unfamiliar topic can result in chaos. Graphics equipment (which increases the richness somewhat) may help to sort out concepts and partition the problem. Computer conferencing (which is less rich than an audio link) may assist with understanding the complexity at the expense of slowing the exchange and introducing an interface which is not necessarily familiar to each person. The richness of full-motion video will make person identification easier, restore the balance between eyes and ears, and should make it possible to handle larger groups.

The richness of the medium employed has other effects. As richness decreases, many persons find it easier to argue for positions they do not support. For this reason, for meetings in which the parties intend to discuss important issues, full-motion video is essential, and even then may not be totally adequate. As richness decreases, the conference participants tend to equalize in importance. Thus, full-motion video allows easy identification of headquarters persons by their manner, style, and dress. In an audio conference, none of the visible marks of rank are available. Attention depends on eloquence, logic, and the ability to assert one's self vocally. A computer conference may even reverse the ranking, for successful participation depends on easy familiarity with terminals and protocols. This is more likely to be found in working-level participants than in management participants.

2.3.3.7.3. User Perceptions. Nevertheless, those who employ teleconferencing facilities report improvements in productivity and the working environment. An important prerequisite appears to be a senior manager determined to use teleconferencing to break the face-to-face meeting habit. When used regularly, teleconferencing results in shorter and more timely meetings. More issues are raised and resolved, decisions can be made on the basis of more correct information and less misinformation, and unproductive time is reduced. The results are said to be greater management visibility, improved communication throughout the organization, improved morale, and increased cooperation.[(6)]

2.3.3.7.4. Inter- and Intraorganization Effects. Most of the applications of teleconferencing reported are within large institutions. This is not surprising since a community of interest is already formed, the resources may be easier to obtain, and the location of the terminals is already established (two separated facilities, for instance). In addition, in most large organizations, there is a sense of belonging which makes employees who have never met not strangers, but unacquainted members of the same *family*. Conditions are such as to heighten the probability that an intraorganization teleconference can be completed over relatively simple facilities (audio-only or augmented audio) to the satisfaction of all parties.

But what of interorganization teleconferencing? Inevitably, this implies more conferences between strangers, so that the richness of the medium must increase in order to have the same probability of satisfying all parties, and more expensive facilities (full audio/video) are required. It also requires compatible terminal equipment (so that one terminal can communicate with any other terminal) and a switched video transmission network (so that the transmission facilities can be shared between many subscribers). Thus, interorganization teleconferencing is more complex. For this reason, it is slower to develop than intraorganization teleconferencing.

2.3.3.7.5. Substitution for Travel. The ubiquitous telephone already replaces a great deal of business and personal travel undertaken to share information and solve problems. In fact, we reach for the telephone without even considering travel as an alternative means of achieving many objectives. However, by expanding our information field, the same instrument stimulates our natural curiosity, so that the simultaneous achievement of greater use of telecommunication, and reduced travel, is probably impossible.

To solve a problem, or share information, a two-way audio conferencing link is as efficient as a face-to-face meeting for acquainted participants. If this is all a meeting achieves, then it can be held over a relatively simple teleconferencing facility, and the travel eliminated. Such is the case, perhaps, for regular meetings of the kind at which project progress is reviewed, or sales information is exchanged. But meeting by teleconferencing is not just a matter of having the same kind of meeting over a communications link. Because the intimate, personal contact is lost, it is less than a face-to-face encounter; it is also much more, for many persons can be present at each end. Thus, teleconferencing could mean that a better outcome is achieved from many meetings because more of those who really know a subject can participate, and new experts can become involved as the topics change. Important decisions need no longer be made by traveling generalists, but can be discussed to any depth within the capability of the resident staff at each terminal location.

Many meetings, however, have complex objectives, and oftentimes the stated purpose for the meeting is only the excuse for accomplishing something else. Some attendees have special reasons for traveling to the meeting site, just as those who wish to communicate may have reasons for staying at home. Some will be official; some may be personal, and public knowledge; and some may be personal, but private. They influence the decision to travel or communicate much more than whether the formal purpose of the meeting can be achieved over a teleconferencing facility.

2.3.3.7.6. Location of Conferencing Facilities. The decision to travel or teleconference will also be influenced by the location of the teleconferencing terminals. Within an organization, greater use of teleconferencing may occur between suburban locations and downtown headquarters, than between suburban locations (one of which may be headquarters). This is particularly true in older, metropolitan complexes where going downtown means traffic snarls, uncertain parking, and danger of vandalism. In this environment, travel problems exceed teleconferencing problems and those working in suburbs could be expected to use teleconferencing as often as the downtown headquarters staff would allow them to. For the same reasons, a downtown public teleconferencing terminal will do little to encourage interorganization teleconferencing unless it is conveniently located, provides adequate supporting facilities (for instance, parking and refreshments), and is easy to use.

2.3.3.7.7. Energy Savings. Specific energy savings due to the use of telecommunication instead of traveling must be calculated for particular cities, number of participants, and mode of travel. Substituting a teleconference for two persons traveling as little as 100 miles by air can result in a significant energy savings. However, while audio conferencing is probably cheaper than travel under almost all circumstances, the operating cost of videoconferencing soon outstrips actual travel cost. Thus, for anything but audio conferencing, saving energy will increase the cost of doing business.

Given the motivation of many individuals to travel (despite ready access to a telephone), and that much of this travel may be outside of the traveler's own organization, it appears unlikely that the availability of teleconferencing facilities will promote a wholesale reduction in travel. For this to occur, the cost of videoconferencing must be substantially reduced or travel must become a lot more difficult and expensive.

In summary, teleconferencing appears to have significant potential for improving productivity and the business environment. in large organizations conditions are conducive to its use. More and more senior managers appear to recognize its potential and are making the decision to break the meeting habit. With this in mind, teleconferencing may come under the influence of market pull in the near future.

2.3.4. Interactive Television

Interactive television is a medium which serves many persons simultaneously, providing a selection of video programming to them, and carrying limited responses in the return direction. It is usual to designate the direction of the video information flow, *downstream,* and the direction of the responses, *upstream.* Downstream information may be broadcast to everyone, distributed point-to-multipoint or point-to-point. Upstream information is usually distributed point-to-point. Interactive television can be implemented using cable television or broadcast television facilities.

2.3.4.1. *Interactive Cable Television*

Cable television (CATV) was first implemented in 1949 in Pennsylvania in an area cut off by mountains from direct reception of television signals. By siting a television antenna on a mountaintop and carrying the received signals down the mountain on cable, it was possible to provide video entertainment to the citizens of the community in the valley. First called Community Antenna Television Service, it became known as Cable Television as suppliers recognized that more than *off-air* signals could be distributed on the cable, making CATV attractive to communities not located in poor reception areas. The result is that almost half the households in the United States are now within reach of cable. Not all cable systems are interactive: only a very few have facilities which transport responses upstream to the *head end,* or video origination point.

2.3.4.2. *Interactive Broadcast Television*

Interactive television can be implemented using broadcast television signals to which the viewer is invited to respond by using the telephone. This technique may also be employed with one-way cable systems. If the voice responses are bridged together so that audio conferencing takes place, interactive television becomes synonymous with aspects of teleconferencing explored earlier.

2.3.4.3. *Applications*

In its simplest form, interactive television may be used for gathering audience reaction to, or opinion of, items presented on the downstream video channel. In these situations only the number of respondents *for* or *against* is tallied. More complex is the use of interactive television to select and order merchandise. Here the items required and the identity of the respondents must be recorded. Even more complex is the use of interactive television to obtain home computer software or games. Most complicated is the use of the medium for information retrieval and computer-aided instruction. These are all activities for which systems have been designed and demonstrated.

Interactive cable television has also been employed to implement services which depend on the return channel capability independent of the downstream video information flow. Thus, interactive CATV systems have been used to provide intrusion and fire-alarm services, meter reading, and hotel/motel status services. These information functions can also be performed by other point-to-point media such as telephone or data communication. (Further discussion of CATV will be found in Section 4.4.5.)

2.3.4.4. *Implementation*

Of some 5000 cable systems which serve close to 20 million households in the United States, less than 100 offer two-way services. Of these, the majority are devoted to alarm and monitor functions. Interactive systems have also been installed in Canada, the United Kingdom, France, Germany, and Japan.

2.3.4.5. *Needs and Motivations*

Interactive television has been the most widely discussed of the new media, yet the least exploited. When conceived in the late 1960s, in the environment of President Lyndon Johnson's "Great Society," it was perceived by social scientists and others to have potential for providing minority access to telecommunication and for providing a medium for the reinforcement of civil rights.[(7)] In spite of some outstanding demonstration projects, these promises have never been attained. Interactive television is an example of a technology seeking an application. It is clearly subject to technology push, not market pull.

2.4. Other Media

The media discussed employ different degrees of richness in the downstream and upstream connections and achieve varying degrees of interaction between users. In Figure 2.10, selected media are shown arranged on this basis. They are divided into two quadrants: one which contains the real-time media, the other which contains record media. In recognition of the fact that some of the latter are electronic, they are also designated store-and-forward media (see Section 4.7.4). In no case does the richness of the upstream channel exceed the richness of the downstream channel.

In Chapter 1, telecommunication media were defined to have functions and contents. Using the correspondence between topology and function, and richness and content, it is possible to form a morphology of telecommunication media. Thus, function depends upon the topology of the downstream and upstream flows. It is related to whether the downstream content is broadcast (one-to-many), sent point-to-multipoint, or sent point-to-point; and whether there is an upstream content, and what is the topology of the flow. One-to-many communication is a downstream function which establishes the location of the receivers. The upstream response will be point-to-multipoint, point-to-point, or none, and

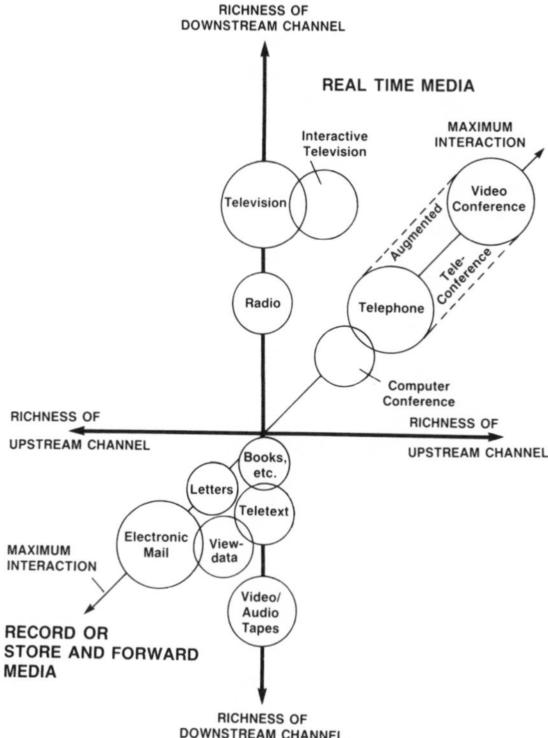

Figure 2.10. Selected media arranged on the basis of richness of upstream and downstream channels and degree of interaction between users.

in order to preserve the dominance relationship between downstream and upstream, the upstream topology will be no more extensive than the downstream topology. In addition, store-and-forward affects both streams, or neither. On the basis of these rules, it is possible to construct the table shown in Figure 2.11 of the topological combinations which characterize the principal telecommunication media.

Content depends upon the type of message carried, that is, whether it be video, audio, or data, or a combination of them. We have called this quality the richness of the medium. To preserve downstream dominance, the richness of the downstream information is greater than, or equal to, the richness of the upstream information. In an interactive mode, the use of upstream video without audio unduly restricts the common communication information space making its contribution marginal. Thus, at the richest level, downstream video and audio are matched by upstream video and audio or responded to by lesser combinations of audio and data (but not video and data). With these rules in mind, it is possible

TOPOLOGY	DOWNSTREAM			STORE AND FORWARD	UPSTREAM		TYPICAL APPLICATIONS	IMPLIED RICHNESS
	ONE-TO-MANY	PT-TO-MPT	PT-TO-PT		PT-TO-MPT	PT-TO-PT		
	INTERACTIVE							
T1	1	0	0	0	0	1	Interactive Radio	R19
T2	0	1	0	0	1	0	Multilocation Teleconference	R19
T3	0	1	0	0	0	1	Video Seminar with Talkback	R7
T4	0	0	1	0	0	1	Telephone Call	R19
	PSEUDO-INTERACTIVE							
T5	1	0	0	1	0	1	Electronic Announcement	R21
T6	0	1	0	1	0	1	Electronic Notice	R21
T7	0	0	1	1	0	0	Electronic Mail	R21
	NO INTERACTION							
T8	1	0	0	0	0	0	TV Broadcast	R23
T9	0	1	0	0	0	0	CATV	R23
T10	0	0	1	0	0	0	Electronic Notice	R28

NOTE: 1 = PRESENT, 0 = ABSENT

Figure 2.11. Topologies which characterize the principal telecommunication media.

RICHNESS	DOWNSTREAM			UPSTREAM			TYPICAL APPLICATION	IMPLIED TOPOLOGY
	VIDEO	AUDIO	DATA	VIDEO	AUDIO	DATA		
	INTERACTIVE							
R1	1	1	1	1	1	1	Videoconference with Augmentation	T4
R2	1	1	1	1	1	0	Videoconference One-Way Augmentation	T4
R3	1	1	1	0	1	1	Videoshopping	T4
R4	1	1	1	0	1	0	CATV with Teletext and Talkback	T3
R5	1	1	1	0	0	1	CATV with Teletext and Alarms	T3
R6	1	1	0	1	1	0	Videotelephone Call	T4
R7	1	1	0	0	1	1	Videoseminar with Talkback and Computer Response	T3
R8	1	1	0	0	1	0	Broadcast TV with Call-In	T1
R9	1	1	0	0	0	1	CATV with Voting	T3
R10	1	0	1	0	1	1	Remote Control	T4
R11	1	0	1	0	1	0	Remote Control	T4
R12	1	0	1	0	0	1	Remote Control	T4
R13	0	1	1	0	1	1	Voice and Data Call	T4
R14	0	1	1	0	1	0	Audioconference with One-Way Augumentation	T2
R15	0	1	1	0	0	1	Audioseminar with Computer Dialog	T3
R16	1	0	0	0	1	1	Remote Control	T4
R17	1	0	0	0	1	0	Remote Control	T4
R18	1	0	0	0	0	1	Remote Control	T4
R19	0	1	0	0	1	0	Telephone Call	T4
R20	0	1	0	0	0	1	Audioseminar with Voting	T3
R21	0	0	1	0	0	1	Computer Conference	T2
	NO INTERACTION							
R22	1	1	1	0	0	0	Broadcast TV with Teletext	T8
R23	1	1	0	0	0	0	Broadcast TV	T8
R24	1	0	1	0	0	0	Videosurveillance with Sensor	T9
R25	0	1	1	0	0	0	Radio and Data Channel	T8
R26	1	0	0	0	0	0	Videosurveillance	T10
R27	0	1	0	0	0	0	Radio Broadcast	T8
R28	0	0	1	0	0	0	Electronic Notice	T9

Figure 2.12. Levels of richness which characterize the major telecommunication media.

to construct the table shown in Figure 2.12 of the richness levels which characterize the principal telecommunication media. In both tables, typical applications are suggested to illustrate the concepts. By no means do they exhaust the possibilities.

Close to 100 combinations of richness and topology can be constructed from these tables. The same level of richness can be associated with several topologies (see R19 in Figure 2.11, for instance). Many of these combinations are only marginally different from others and some are obviously more costly than others. Whether they are used depends on more parameters than that they are technical options which can be implemented. Further, there are opportunities to use two (or more) of these simple combinations together. Thus, it is common to see discussions between persons located in different cities over network TV (or CATV). This involves one set of persons (the discussion group) who employ a multipoint videoconference combination (T2, R6) and a one-to-many (T8, R6) or point-to-multipoint (T9, R6) connection to the television audience. The opportunities for special telecommunication arrangements probably outdistance the requirements for them, and the average customer's ability to pay for them. What their *messages* might be is difficult to estimate.

3

Technology

Telecommunication is achieved through electrical and optical signals carried in some fashion from sender to receiver. To understand the procedures involved, the equipment employed, and the effects new technologies are likely to have on them requires an appreciation of the basic engineering science associated with telecommunication facilities. This includes the properties of voice, data, and video signals; the elements of transmission and digital switching; concepts in traffic theory, software, and protocols; and the principles of operation as well as the methods of construction of electronic and optical components and devices. These subjects are organized into three sections devoted to signals, control, and microstructures.

A. SIGNALS

In this section we discuss the basic properties of voice, data, and video signals and the fundamentals of transmission and switching facilities.

3.1. Basic Concepts

Some concepts are fundamental to telecommunication equipment and facilities: they include the different nature of analog and digital signals, the significance of time and frequency analysis, the definition of power ratio, the effect of noise, and the action of modulation.

3.1.1. Analog and Digital Signals

An *analog* signal is well defined and continuous. It may assume positive or negative values, but at all times the value *now* is related to the value that *was,* and the value *to be,* by a smooth, continuous variation. In mathematical parlance, the rate of change of value with time (dV/dt) is always finite. In contrast, a *digital* signal assumes discrete states only. By far the most popular class of digital signals are *binary* valued, that is, they exist in two states, usually denoted by 0 and 1. Changes of state occur instantaneously and the rate of change is infinite. Thus, digital signals are discrete and discontinuous. While the concept of instantaneous change is a simple matter for the mathematician, no natural process exactly fulfills this condition. Nevertheless, the idea of two classes of signals (analog and digital) is important. As we shall see later, analog signals are better suited to the transmission of information, and digital signals are better suited to the manipulation of information.

Analog signals can be converted to digital signals, and vice versa. Conversion is achieved by sampling the amplitude of the analog signal at regular intervals and comparing these amplitudes to a series of predetermined levels established in a converter circuit. The digital equivalent of each analog sample is the binary number which denotes that level which is just exceeded. The digital equivalent of the analog signal is the series of these numbers which describe successive samples. Reconstruction consists of generating a series of pulses whose amplitudes are equal to the same fractional levels of the maximum voltage employed in the reconstructing circuit and applying them to a suitable filter. The resulting signal will be a good approximation to the original signal provided the sampling rate was at least twice as frequent as the highest frequency present in the original signal. This relationship is known as the (Nyquist) *sampling theorem* and the rate is known as the Nyquist rate. The restored signal will deviate from the original by an amount related to the approximation made in matching the analog samples to the fractional levels in the digitizing circuit. The more levels employed, the more faithfully will the restored signal represent the original.

3.1.2. Time and Frequency Analysis

The signals encountered in telecommunication are too complicated to be used directly in the analysis and description of systems. Fortunately, they can be represented by sets of *sine* and *cosine* signals, a result demonstrated by Jean Baptiste Fourier. In the early 19th century, he showed that repetitive, complex functions can be considered to be made up of sets of harmonically related *sinusoidal* functions which we now call *Fourier series*. Provided the system is *linear,* that is to say there is an unchanging relationship between the magnitude of output and input for all values of the input waveform applied to the system, the response to a complex, repetitive signal is the same as the sum of the responses to the sinusoidal signals which are the components of the Fourier series representing the complex signal. This result is known as the *principle of superposition*. In combination with Fourier series and similar transformations, it forms the basis for the analysis of modern telecommunication systems.

Sinusoidal functions occur naturally. The motion of a simple pendulum is

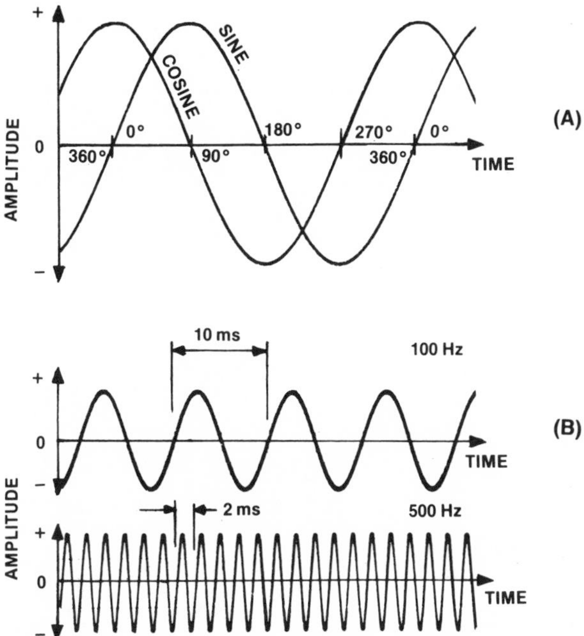

Figure 3.1. Important properties of sinusodial functions. In (A), the relationship between (signal) amplitude and time is shown for both sine and cosine waves (functions). A single cycle is divided into 360 degrees. These angles are known as the *phase* of the signal, and are repeated each cycle. Sine and cosine functions are displaced by 90 degrees from each other (the cosine function *leads* the sine function by 90 degrees and they are said to differ by a phase angle of 90 degrees). In (B), two sine waves are shown. In one, a cycle occupies 10 ms (thus the values of the wave are repeated 100 times each second): it is said to have a *frequency* of 100 cycles per second (or 100 hertz, or 100 Hz). The other has a frequency of 500 Hz, and oscillates five times as rapidly.

sinusoidal, and the stable solutions of the differential equations describing the behavior of many physical systems are sinusoidal functions. They have three important parameters—frequency, phase, and amplitude—which are illustrated in Figure 3.1.

A repetitive, complex signal $S(\omega,t)$, which comprises time-varying energy distributed in frequency and time, can be represented by the set of sinusoidal functions given by

$$\begin{aligned} S(\omega,t) = {} & a_0 + a_1 \sin(\omega t + \phi) + a_2 \sin(2\omega t + \phi_2) \\ & + \cdots + a_n \sin(n\omega t + \phi_n) + b_1 \cos(\omega t + \psi_1) \\ & + b_2 \cos(2\omega t + \psi_2) + \cdots + b_n \cos(n\omega t + \psi_n) \end{aligned} \tag{3.1}$$

where $\omega = 2\pi f$.

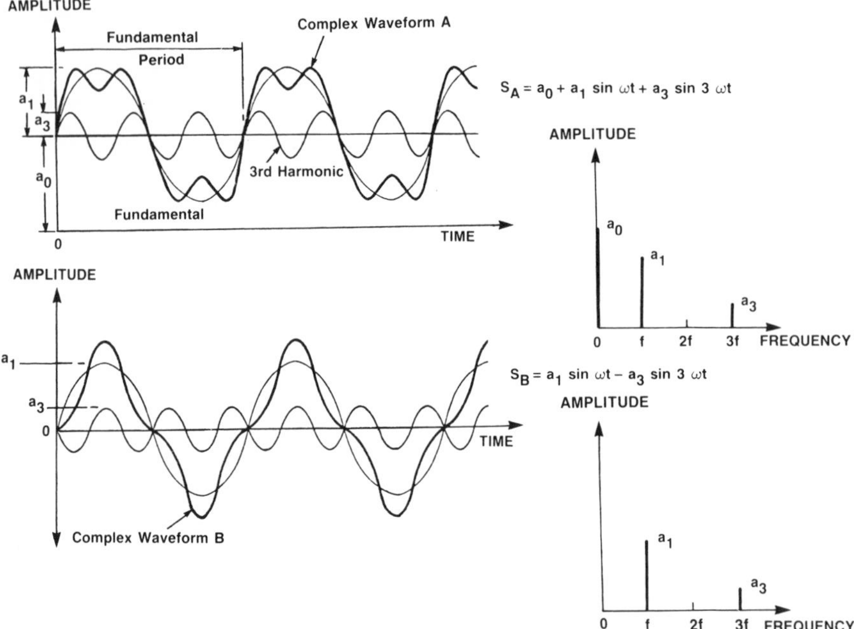

Figure 3.2. Complex waveforms A and B consist of fundamental and third harmonic components. In addition, waveform A is displaced from the zero amplitude axis by a constant term, producing a frequency spectrum which has energy at zero frequency.

This Fourier series contains a constant term a_0, and terms $a_1, a_2, \ldots, a_n$ and $b_1, b_2, \ldots, b_n$ which define the peak amplitudes of the sinusoidal components. Frequency, f, is contained in the arguments of the sinusoidal terms, as is time, t, and phase, ϕ and ψ. The frequency of each term is harmonically related to the frequency of the *fundamental* terms $a_1 \sin \omega t$ and $b_1 \cos \omega t$. The higher-frequency terms are known as *harmonics* (second, third, etc.).

Examples of two complex waveforms are given in Figure 3.2. Complex waveform A consists of three components: a constant term a_0, a fundamental component $a_1 \sin \omega t$, and a third harmonic term $a_3 \sin 3\omega t$. The frequency spectrum consists of three components: a_0 at zero frequency, a_1 at the fundamental frequency, and a_3 at three times the fundamental frequency. The waveform is displaced to the positive side of zero amplitude by the constant term a_0.

Complex waveform B consists of two components: a fundamental term $a_1 \sin \omega t$ and a third harmonic term $-a_3 \sin 3\omega t$. The third harmonic is 180° (or π radians) out of phase with the third harmonic term in waveform A. Unlike waveform A, it is symmetrical about zero amplitude. The frequency spectrum consists of two components: a_1 at frequency *f*, and a_3 at frequency 3*f*. There is no zero-frequency component.

As we shall see later, the presence of a zero-frequency component (constant term) can be a handicap in certain applications. To transmit waveform B, a *channel* which will pass frequencies *f* and 3*f* is required. The *bandwidth* of the signal, that is, the smallest frequency interval which contains all of the energy associated with the signal, is 2*f* Hz. In the case of waveform A, the bandwidth is 3*f* Hz and the channel must transmit energy at zero frequency. For most transmission channels, this is impossible because of the characteristics of the devices employed or the modulation techniques used.

But what if the complex signal is not exactly repetitive? What if the signal is not repetitive at all? Then the amplitude versus frequency representation ceases to be a series of discrete lines and the intervening space is filled with energy at all frequencies. A well-known example is provided by a lightning stroke. It consists of a sharp release of energy of very great amplitude and very short duration which radiates across a wide frequency range producing characteristic interference to radio and television signals.

3.1.3. Noise

The performance of telecommunication equipment and systems is limited by the level of interfering signals which mask or distort the signal carrying the information of interest. These unwanted signals may be due to other equipment *spilling* over into the channel, to the effects of processing the signal, to electron activity in the front end of the receiving equipment, and to atmospheric disturbances such as lightning strokes, sunspots, etc. All of these signals are called *noise*. Because they appear neither deterministic, nor ordered, noise signals are usually discussed in terms of the average power present. An important parameter is the signal-to-noise ratio, usually written S/N.

In electronic devices, noise can be attributed to two major processes, thermal noise and shot noise. *Thermal* noise is produced by the random motion of electrons in the signal-carrying medium. The intensity of their motion increases with temperature, and is zero only at a temperature of absolute zero. *Shot* noise is produced by the discrete nature of electron motion. While the flow of electric current can be characterized as a continuous process in metal conductors, in other materials it consists of electrons darting from potential barrier to potential barrier within the atomic structure of the material supporting conduction. The fluctuations produced by these movements produce a rapid, random variation about the average current, which can interfere with the desired signal.

To the device developer and materials physicist it is important to differentiate and characterize thermal and shot noise processes in detail. For the telecommunication engineer it is often adequate to represent the sum of all sources of noise as a random process (stochastic process) with a constant power spectrum density—that is, a constant average noise energy density is present across the frequency band of interest. This approximation is called *white* noise, where white is used in the same sense that white light contains all colors (frequencies). For very wide frequency ranges, the white noise approximation may not be adequate. Instead, a power spectral density in which average noise power varies with the inverse of frequency may be used. Called $1/f$ noise, it contains an increasing amount of energy at lower frequencies. Observations of $1/f$ noise suggest a correlation between present values of noise and past events[(1)]; that is, it depends on the use the system has already sustained.

3.1.4. The Decibel (dB)

An important measure in telecommunication systems is the ratio of powers. It is often expresed in *decibels* (or dB), a dimensionless quantity representing 10 times the logarithm (to base 10) of the ratio of two powers; thus, the ratio (D) of P_1 and P_2 in decibels is given by

$$D = 10 \log_{10} (P_2/P_1) \text{ dB} \tag{3.2}$$

To calculate the overall gain, or loss, of a system with several sections, it is only necessary to add up the ratios (in dB) of output power to input power for each section. Power levels existing at significant points in a system can be expressed in dB in relation to an arbitrary power level, or to the power at one point in the system (often called the zero-level point). Since power is proportional to the square of amplitude (A, of voltage or current), the use of decibels can be extended to relate ratios of voltages or currents. In this case

$$D = 10 \log_{10} (A_2^2/A_1^2) = 20 \log_{10} (A_2/A_1) \text{ dB} \tag{3.3}$$

Obviously, it is important to state whether powers or amplitudes are being compared when reporting dB levels. In Figure 3.3 values are given to illustrate the relation between decibels and powers related to a reference level of 1 milliwatt (1 watt divided by 1000, 1 mW) and between decibels and the ratio of two voltages.

3.1.5. Analog Modulation

Two sine waves of differing frequencies can be mixed together in a nonlinear device (i.e., a device in which a constant ratio between output and input does not hold for all values of input signal applied to it) to produce a composite signal in which the lower-frequency wave (called the signal) is *modulated* on the higher-

(A) ILLUSTRATING THE RELATION BETWEEN DECIBELS (dB) AND POWER RATIO (P_2/P_1)

$$dB = 10 \log_{10} (P_2/P_1)$$

FOR P_2/P_1 =	1000	100	40	20	10	4	2	1	0.5	0.25	0.1	0.01
dB =	30	20	16	13	10	6	3	0	−3	−6	−10	−20

(B) ILLUSTRATING THE RELATION BETWEEN POWER AND A REFERENCE LEVEL OF 1 mW. THE NOTATION dBm IS USED FOR THIS SPECIFIC RATIO.

$$dBm = 10 (3 + \log_{10} P_2)$$

FOR P_2 =	1W	1 mW	1 μW
dBm =	30	0	−30

(C) ILLUSTRATING THE RELATION BETWEEN DECIBELS AND VOLTAGE RATIO (V_2/V_1)

$$dB = 20 \log_{10} (V_2/V_1)$$

FOR V_2/V_1 =	1000	100	40	20	10	4	2	1	0.5	0.25	0.1	0.01
dB =	60	40	32	26	20	12	6	0	−6	−12	−20	−40

Figure 3.3. Decibel values.

frequency wave (called the *carrier*). Using this technique, audio or video signals can be impressed on broadcast carriers. The bandwidth of the modulated carrier is determined by the bandwidth of the audio or video signal (called the *baseband*) and the modulation technique employed. Upon arrival at a receiver, the lower-frequency signal is separated from the carrier and retrieved in a process called *demodulation*.

Three basic modulation schemes are possible: amplitude, frequency, and phase. The relationship between amplitude and time for these techniques is shown in Figure 3.4. In *amplitude* modulation, the amplitude of the higher-frequency carrier wave is modulated by the amplitude of the lower-frequency signal wave. If the carrier has a frequency of f_c, and the signal has a frequency of f_s, the result is a set of three frequencies: $(f_c - f_s)$, f_c, and $(f_c + f_s)$. The effect of amplitude modulation is to reflect the signal frequency symmetrically about the

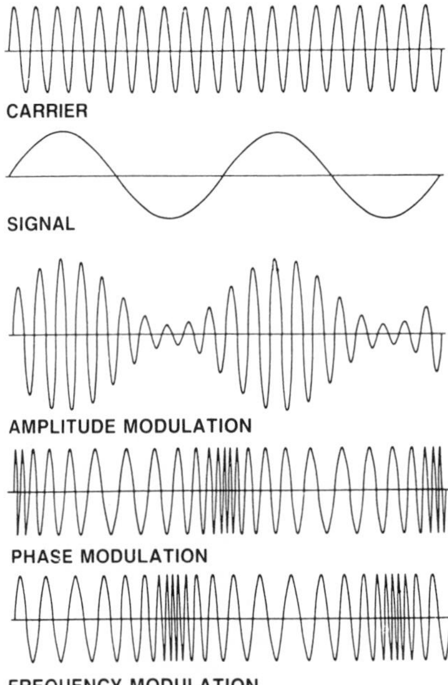

Figure 3.4. Amplitude, phase, and frequency modulation.

carrier frequency. This is known as *double-sideband amplitude modulation* (DSB-AM). The upper and lower sidebands contain the same information: if one is eliminated, the arrangement is known as *single-sideband amplitude modulation* (SSB-AM). The result is a decrease in the total bandwidth of the modulated signal. Besides suppressing one sideband, it is possible to suppress the carrier signal (SSB-SC) also, significantly reducing the energy contained in the composite signal, but making demodulation more difficult. A better approach is to partially suppress one sideband so that an attenuated carrier is transmitted. The presence of the carrier allows effective demodulation. Known as *vestigial-sideband amplitude modulation* (VSB-AM), this technique provides virtually the same quality demodulated signal as is given by DSB-AM, but uses less frequency spectrum and requires less power. These modulation techniques are illustrated in Figure 3.5.

Phase modulation (PM) and *frequency modulation* (FM) are particular cases of *angle modulation*. In PM, the instantaneous phase deviation from the undisturbed carrier is proportional to the modulating signal amplitude. In FM, the instantaneous frequency deviation is proportional to the modulating signal amplitude. Examination of Figure 3.4 shows that the modulated waves from FM and PM have similar shapes: they both contain varying frequency and phase components. However, the relationship betwen the points of maximum frequency of the modulated wave and the peaks of the modulating signal are different. The bandwidth required for PM and FM cannot be given precisely. It depends on the waveform of the modulating signal and the fidelity with which the demodulated signal must match the modulating signal. An estimate can be obtained by taking twice the sum of the highest frequency in the modulating signal and the

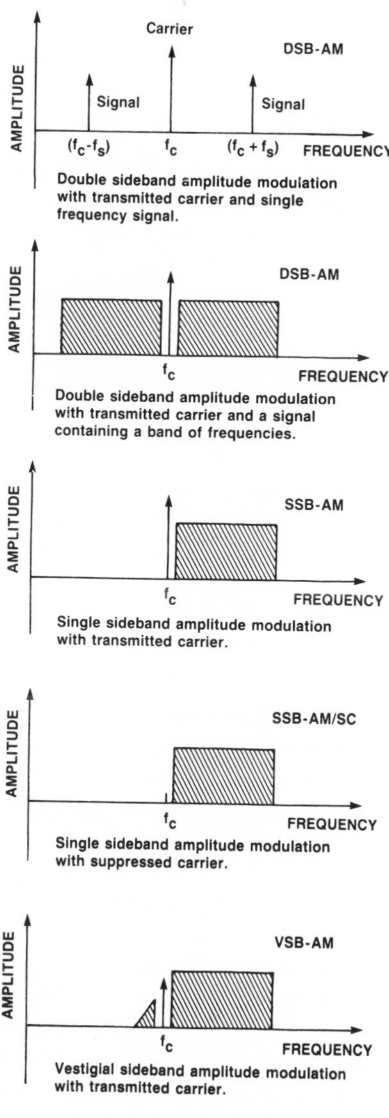

Figure 3.5. Modulation produces sidebands. They can be suppressed, or shaped in different ways.

peak frequency deviation. Digital modulation, which may employ PM or FM, is described in Section 3.3.4 of this chapter.

3.2. Speech/Sound Signals

Speech is the most important personal communication medium. It gives rise to complex sounds which are converted to equally complicated analog signals. In turn, they can be converted to digital signals and encoded in various ways.

3.2.1. Natural Voice

Speech, a major input to many communications systems, consists of strings of basic sounds, called *phonemes*, uttered at rates up to ten per second (10/s) to form words; their existence can be shown by recording a running energy spectrum (or sonogram) of a series of spoken words. Such a record is shown in Figure 3.6. Identifiable sounds are formed by resonances in the vocal tract which are excited by bursts of air originating in the voice box and modulated by vocal cords. Careful inspection of Figure 3.6 reveals a line structure which represents the vocal cord vibration period. The exact nature of the vocal cord vibrations greatly influences the production of speech. Coupled with the cavities which make up each person's vocal tract, they produce the unique sounds that unmistakably distinguish each of us. Determination of the resonant frequencies of the combination (formant frequencies) is important in many aspects of speech processing, including speaker identification, speech recognition, and speech synthesis. The resonances of the vocal tract give rise to the dark stripes running through the utterances of Figure 3.6. Figure 3.6 also shows the same utterances as they appear at the ear of the listening party after transmission through the speaker's telephone, the speaker's local connection, the switch, the listener's local connection, and his telephone. The system has obviously distorted these utterances, and may have added interfering signals (noise), yet, thanks to the versatility of the listener's brain, they are intelligible to him. Frequency, phase, and amplitude are manifest in different ways.

3.2.1.1. Frequency

To our ears, frequency is manifest as *pitch*. Just as music requires a range of frequencies to achieve near-lifelike reproduction, so do other messages. Each message requires a specific *bandwidth*. The human voice may generate frequencies from 100 Hz to 7 kHz (a bandwidth of 6900 Hz), with most of the energy contained between 200 Hz and 1 kHz. A full orchestra may produce frequencies between 20 Hz and 20 kHz (a bandwidth of nearly 20 kHz.). In telephone networks, voice is limited to the range from approximately 300 Hz to 3.4 kHz. Broadcast music is limited to 5 kHz (AM) or 15 kHz (FM).

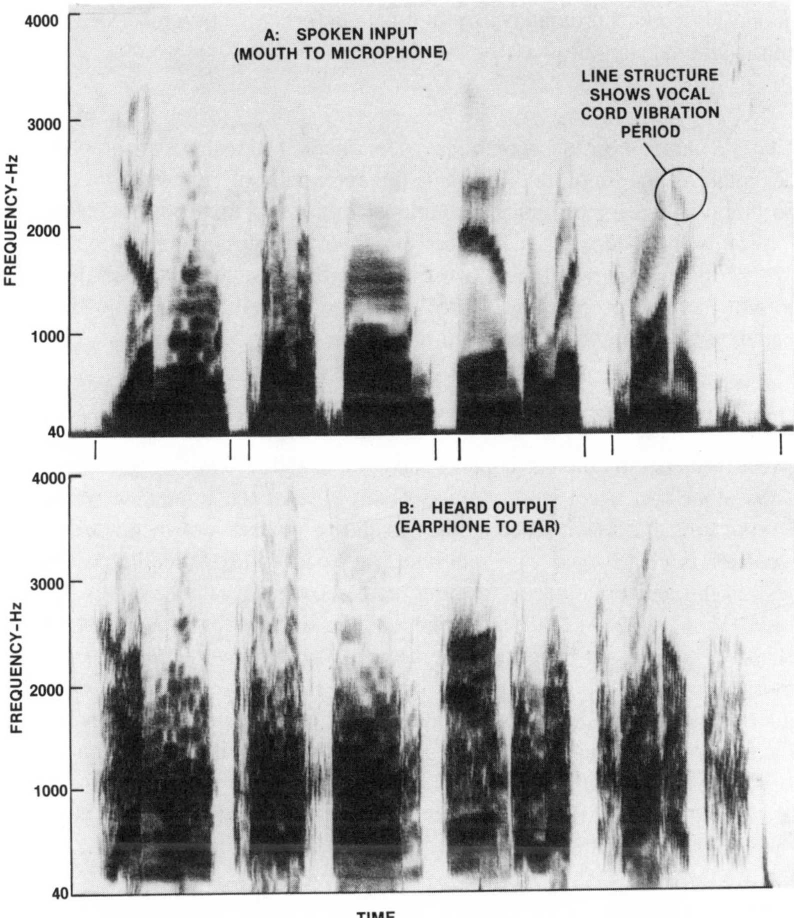

Figure 3.6. Sonogram of words spoken into telephone microphone (A) and heard from telephone earphone (B). The resonances of the vocal tract appear as dark stripes which are horizontal for a steady vowel, and which bend up or down during transitions between consonant and a vowel, or from one vowel to another. Much of the fine detail is destroyed—or distorted—by passage through the local network. Nevertheless, the sound is easily understood by the recipient.

3.2.1.2. Phase

To our ears, phase is manifest as intelligibility. Changing the phase of a signal (but not frequency or amplitude) changes the location of events—and can destroy much of the correlation between signals.[2] This effect is manifest in the waveforms of Figure 3.6. Peaks and troughs may be smoothed, thereby reducing contrast, and smearing the detail of the input to produce the output signal shown in Figure 3.6B. Certainly some of the differences between sonograms A and B in this figure are due to phase changes caused by the characteristics of the telephone network. Fortunately, the brain is a forgiving processor which can accommodate substantial phase degradation.

3.2.1.3. Amplitude

To our ears, amplitude is manifest as loudness. The response is proportional to the square of the amplitude. Further, the perception of change of loudness is proportional to the logarithm of the ratio of initial and final powers. Doubling the power will produce a specific perception of loudness change, doubling it again will increase perception to two times the initial state, and doubling it yet again will increase perception to three times the original, etc. This logarithmic dependence is the function discussed in Section 3.1.4.

3.2.1.4. Talkspurts

In a conversation the participants share the task of talking. First one speaks, then the other. On average, each talks about 40% of the time. The remaining 20% represents the period when neither is talking—pauses between words when the speaker is clearly going to continue, and pauses for reflection when one ceases talking and the other is forming his response. Occasionally, the talking activities will overlap as one interrupts the other. The periods in which the speakers are generating continuous vocal energy are known as *talkspurts*. Considered from this perspective, voice messages consist of a series of bursts of sound energy of varying duration interspersed with periods of silence. This characteristic can be put to good use in order to share facilities between several talkers (see Sections 4.1.1.5.2 and 4.1.1.6). Successful implementation depends on being able to detect the cessation and onset of vocalization without chopping the beginnings or endings of words. *Silence detection* must be done almost perfectly to prevent disturbance of the conversation.

3.2.2. Digital Voice

Speech produces a continuous analog signal which can be converted to a digital signal by sampling the signal amplitude at a rate of 8000 samples per second to produce a train of pulses from which signal components up to 4 kHz can be retrieved. The accuracy of the reconstruction depends on the number of discrete levels available for matching to the samples. In turn, this affects the number of bits used to describe the levels. Choosing the number is a compromise between fidelity of reproduction (which requires more levels or bits), the cost of transmission facilities (a lower bit rate means less expensive facilities), and the ability to resist degradation when the signal is encoded and decoded a number of times as it is handled by different facilities in a long-distance call (which requires more bits). Bearing these factors in mind, telephone administrations have adopted a coding called *pulse code modulation* (PCM) in which each sample amplitude is quantized in $2^8 = 256$ levels. Digital voice is carried at a rate of

64 kb/s (8 bits in each of 8000 samples per second) in the world's telephone systems.

3.2.2.1. Companding

In order to accommodate different speaking levels, lower volume signals are matched to relatively more steps in the analog-to-digital conversion than higher-level signals. This results in compression and expansion of the signal *(companding)*. The degree of companding is chosen so that the signal-to-distortion ratio is about constant for a wide variety of signal levels. In the United States, companding is accomplished according to a modified logarithmic relationship known as the μ-law. Elsewhere in the world, A-law companding is employed. In this embodiment, a linear segment is substituted for the logarithmic curve at small signal levels.

3.2.2.2. Coding

As we have already discussed, digital voice which must stand the exigencies of worldwide service employs 64 kb/s companded PCM with 256 levels (8 bits) and a sampling rate of 8000/s. For limited distance service, 128 levels (7 bits, 56 kb/s) will suffice. Other coding schemes can be employed which result in lower bit rates. They increase the complexity of the processing required and may reduce perceived quality. Lower bit-rate coding is of importance in long-distance transmission systems, including satellite systems, since it allows more customers to use a given facility simultaneously. Also, it can be important in broadcast radio systems, such as mobile telephone systems, where the lower bit rate may allow digital voice messages to be carried in the same bandwidth as is now employed with analog voice. (Further discussion of the significance of reduced bit rate voice is reserved for Section 5.1.6.)

The *quality* of the voice signal delivered by various coding schemes is a matter of human judgment. While not measurable in any absolute way, several levels have been defined which are recognizable to most persons. In decreasing order of fidelity, they are commentary quality, toll quality, communication quality, and synthetic quality. *Commentary quality* voice contains the full range of voice frequencies and is suited to radio broadcasting (particularly FM broadcasting) and television sound. It requires approximately twice the bandwidth that telephone voice provides. Such performance can be achieved by 8-bit companded PCM with a sampling rate of 16,000 per second. As a practical matter, it can be implemented over 2 $\times$ 64 kb/s voice channels.

Toll quality voice is generally considered to contain voice frequencies from 300 Hz to 3.4 kHz, have a signal-to-noise ratio of greater than 30 db, and a harmonic content of no more than 3%. It can be implemented as 64 kb/s, 8 bit companded PCM on good quality telephone circuits. Other coding techniques can be used to produce comparable quality at lower bit rates.

Communication quality voice may contain fewer frequencies and more noise than toll quality. It is not necessarily intelligible to the casual listener and is intended for use by professional communicators such as airline pilots and vehicle fleet operators—persons who have short, terse messages to exchange, and who naturally use mnemonics to ensure that vital information (such as flight number, or altitude) is understood.

Synthetic quality voice consists of sounds generated by an electronic analog of the human speech production mechanism in response to information developed by analyzing the input speech. Present analogs produce an artificial voice sound which lacks the characteristics of the speaker. Nevertheless, it may be more intelligible to the casual listener than some communications quality signals.

Speech coding techniques are of two kinds: *waveform coding* and *source coding*. In waveform coding, speech is reduced to a string of bits by operating on the signal in some way. In source coding the speech production mechanism is modeled.

3.2.2.2.1. Waveform Coding. Waveform coding techniques can be divided into two categories: differential techniques and frequency domain techniques.[(3)]

Differential Coding. A powerful technique employed to reduce the bit rate exploits the correlation between adjacent signal samples. Since the signal does not vary rapidly from sample to sample, the difference between samples represents a lower bit rate than that of the signal itself. At the receiver, these differences can be integrated to form an estimate of the input signal. The exact manner in which the estimate is formed (predicted) leads to several differential coding schemes. Only the most important will be mentioned. They are *delta modulation* (DM), *adaptive DM* (ADM), *differential PCM* (DPCM), and *adaptive DPCM* (ADPCM).

Delta modulation is a differential coding technique in which the sampling rate is many times the Nyquist rate for the input signal. (The ratio of the sampling frequency to twice the Nyquist rate is known as the oversampling index. Practical values of the index are between 2 and 32.) The result is that the correlation between samples is very high, and the difference is very small. A one-bit quantizer is employed and the transmitted signal is a series of $+1$'s and -1's which represent a positive or negative difference between successive samples. Integrating these signals produces a good approximation to the input signal provided the step size is greater than the difference between the input signal samples in the region of maximum change. If it is not, the estimated signal will fall behind the input signal, producing a condition known as *slope overload*. As an alternative to increasing the sampling rate to compensate for exceeding the step size, step size adaptation can be employed. This technique is known as adaptive DM. Through its use, toll quality voice can be achieved at 40 kb/s and communications quality around 24 kb/s.

Differential PCM is a coding technique in which the difference quantizer has two or more levels. Like DM, an adaptation algorithm can be employed to vary the quantizer levels in accordance with the rate of change of the input signal, producing adaptive DPCM. These changes ae used to improve both dynamic range and signal-to-noise ratio. ADPCM produces toll quality voice at 32 kb/s and communications quality around 16 kb/s.

Frequency Domain Coding. The differential coding techniques described above operate in the time domain on the complete input signal. Frequency domain techniques divide the signal into a number of frequency bands and code each separately. Because speech is not usually energetic across the entire frequency spectrum simultaneously, it is possible to dynamically allocate more bits (number of levels) to those bands in which the activity is significant at the expense of those bands which are quiescent. Two techniques of this sort are subband coding (SBC) and adaptive transform coding (ATC).

In *subband coding,* the input signal frequency band is divided into several subbands (typically 4 or 8). Each subband is encoded using an adaptive step-size PCM (APCM) technique. SBC produces toll quality voice at 24 kb/s and communications quality around 9.6 kb/s. In *adaptive transform coding,* the input signal is divided into short time blocks (frames) which are transformed into the frequency domain using techniques based on Fourier series. The result is a set of transform coefficients which describe each frame. These coefficients are quantized into numbers of levels based on the expected spectral levels of the input signal. The number of bits for any quantizer, and the sum of all the bits per block are limited to a predetermined maximum. ATC produces toll quality voice at 16 kb/s, and communications quality around 7 kb/s.

3.2.2.2.2. Source Coding. By transmitting a description of the speech and emulating the mechanism of the vocal tract in the receiver, it is possible to produce artificial speech at greatly reduced bit rates. Known as *vocoding,* or voice coding, in early systems (called spectrum channel vocoders), the speech signal was divided into several separate frequency bands (e.g., 16 bands of 200 Hz each). The magnitude of the energy in each band was transmitted to the receiver together with information on the presence or absence of pitch (voiced or unvoiced sounds). At the receiver, voiced or unvoiced energy was applied to a second set of filters to produce recognizable speech. Later equipment (called formant vocoders) made use of format amplitudes and frequencies. Vocoders produce intelligible speech of communication/synthetic quality using transmission rates around 2 kb/s.

In modern systems, speech analysis and synthesis are performed using *linear prediction models* in which successive samples of speech are compared with previous samples to obtain a set of *linear prediction coefficients* (LPCs) which can be used to reproduce the sound. In effect the transmitter *listens* to the input speech, calculates LPC strings, and transmits them to the receiver which regenerates the sound using an artificial vocal tract.

In Section 3.2.1, it was noted that speech is composed of basic sounds, called phonemes, which are uttered at rates up to ten times a second. Reducing voice to a string of codes representing these sounds could produce a signal of the order of a few hundred bits per second. To achieve this rate, significant analysis of the signal is required. Even with today's substantial computers, this cannot be performed in real time.

3.2.2.2.3. Delay. In order to be processed and manipulated, the voice signal must remain within the encoding/decoding equipment for a finite time. The result is that the output train is displaced from the input train by a time delay. For 64 kb/s PCM, this delay is the same as the sampling interval, i.e., 1/8000 s, or 125 μs. For ADM and ADPCM, the delay may be 250 μs, for SBC it may be 20 ms, for ATC over 100 ms, and for vocoders up to one-half a second (500 ms). Any delay greater than 10–20 ms will cause discernible echos in telephone connections. For this reason echo control must be employed with the more complex coding schemes. This topic is discussed further in Section 4.1.1.4.

3.3. Data/Binary Signals

Binary signals are most commonly encountered as the output from computers and data processing devices. As a stream of rectangular pulses, they occupy a large bandwidth and contain low-frequency energy. Coding and pulse shaping are used to restrict the bandwidth of digital signals transmitted in telecommunication systems.

3.3.1. Line Coding

What if we wish to transmit data signals over telephone circuits—which do not pass low frequencies—or employ modulation techniques which require a narrow, clear channel for transmission of the carrier frequency? It is necessary to employ representations for the binary symbols which reduce the energy at low frequencies. They are called *line codes*. In simple terms, the objective is to achieve an equal number of positive and negative pulses over a relatively short time interval so that the constant term in the Fourier series representing these waveforms becomes zero.

3.3.1.1. Binary Codes

One way to achieve the desired balance is to represent 0's and 1's by different polarity pulses, and use a substitution technique to maintain an equal number of 0's and 1's. This can be done by replacing segments of the bit stream by longer blocks. An illustrative example is given in Figure 3.7. Segments of four binary symbols (4B) taken from the message stream are replaced by one of the 18 combinations of six binary symbols (6B) in which there are three 0's

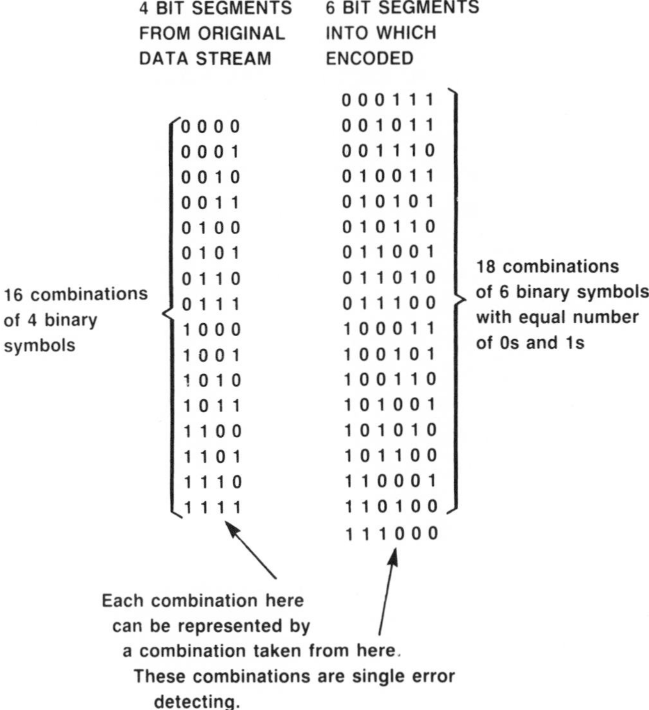

Figure 3.7. Substitution coding to achieve equal 0's and 1's.

and three 1's. The resulting code is known as 4B6B. Block codes of this type increase the line symbol rate. In the example, four message bits are replaced by six bits, producing a 50% increase in required line speed. Less drastic increases will be produced by replacing 6-bit message blocks by 8-bit codes (33% increase), 8-bit message blocks by 10-bit codes (24% increase), etc. The penalty is the need to manipulate larger strings of symbols. Using these block codes makes error detection easy (and eliminates the need for the use of a parity bit). If the received signal block does not match one of the codes in use, an error has occurred. An alternative is to substitute blocks of ternary code for blocks of binary message symbols. The sixteen, 4-bit blocks of message symbols can be replaced by 16 of the 27 possible arrangements of 3 ternary symbols. This coding is designated 4B3T (4 binary-to-3 ternary). Another technique is called *partial response encoding*. Zeros and ones are encoded as sets of two pulses: a positive followed by a negative, or a negative followed by a positive. The second pulse of one symbol occurs at the same time as the first pulse of the next symbol. The result is a series of positive, zero, and negative levels.

3.3.1.2. Bipolar Codes

It is relatively easy to use a three-level code to represent the binary symbols, thereby preserving balance and reducing line rate. In the simplest bipolar coding, 0's are signified by zero level, while 1's are alternately positive and negative. Called *alternate mark inversion* (AMI) coding, the symbols are often referred to as marks (1's), spaces (0's) and negative, or inverted, marks (also 1's). An example of this code is shown in Figure 3.8. A shortcoming is that when no message signal is present, no line signal is generated. In many circumstances, long silences are undesirable because they allow the receiving circuits to fall out of synchronism with the transmitter (see Section 4.1.1.1.2). Accordingly, in some systems, when a string of n zeros have occurred, the $n + 1$ zero is replaced by a mark of the same polarity as the last message mark. This clearly identifies the additional mark as a substituted signal. A further refinement provides that in long strings of 0's, each added mark alternates in polarity. In this way balance and synchronization are maintained through long silent periods.

Other coding schemes are used. Thus, WAL2 employs what is effectively a 4-bit representation. A zero is represented by 1001 and one is represented by

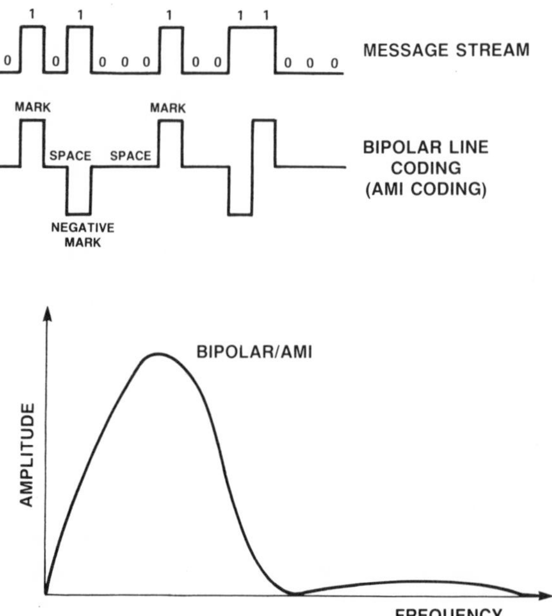

Figure 3.8. Alternate mark inversion coding of a binary message stream. The frequency spectrum shows a significant decrease in energy at low frequencies.

0110. Each symbol is symmetrical so that, when implemented as $+1$ and -1, there is no zero-frequency component, and a string of zeros or ones results in an alternating signal. The data stream can be thought of as the combination of two streams of square pulses which are displaced by 180°. The phase of the data changes by 180° when the symbols change. Another code, known as *coded mark inversion* (CMI), represents zero by 01 and one by 00 or 11 alternately. The data stream can be thought of as the combination of two streams of square pulses, one of which is twice the rate of the other.

3.3.2. Bandwidth

As a practical matter, the bandwidth of a data signal is defined as the band of frequencies which just contains 99% (-20 dB) of the signal energy, *or* the band of frequencies in which the energy had fallen to a specific fraction of the maximum (e.g., -3 db, -50 dB, etc.). Much of the extension of the frequency spectrum of a square pulse can be overcome by shaping the signal pulse to eliminate the higher-frequency components. The selection of the *best* pulse shape for a particular digital signal is a complex trade-off between energy distribution, bandwidth, and intersymbol interference.

3.3.3. Packets

Data communication, the exchange of digital information between machines, can occur over a wide range of message lengths and bit rates. Unlike voice communication in which message lengths may vary but the speed must equal or exceed the sampling rate multiplied by the number of bits used to represent the range of amplitudes, there is no lower limit to data speed. In fact, for many applications there are no requirements for data to be received in a given time, just that it arrive not too long after being sent. In addition, many data messages consist of only a few bits of information. The result is that data communication can be needlessly expensive if a complete circuit is dedicated to each connection. Realization of the bursty nature of data messages, and that the peak data transfer rate between two machines may be one or two orders of magnitude greater than the average rate, has given rise to the concept of *packets*—of collecting data in limited strings which are transmitted at a rate compatible with the medium employed. Packet communication between two machines takes place only when a packet of information has been accumulated and is ready for transmission. During the remainder of the time, the channel and associated telecommunication facilities may be used for packets traveling between other machines.

A packet may be of any size. Packets in use today range from a few bytes long (1 byte = 8 bits) to a hundred or a thousand bytes in length. The longer the packet, the more time required to assemble the contents, and the greater the delay introduced in the transfer of information. The use of packets has given rise to new concepts of switching and transmission which are explored in Section 4.7.4.

3.3.4. Digital Modulation

Because they involve discrete jumps from one state to another, the modulating action with *digital* signals is generally described as *keying*. The simplest technique, known as *amplitude-shift keying* (ASK), is to modulate the carrier with the binary signal to produce a DSB-AM signal. Alternatively, a combination of amplitude and phase can be used to produce a modulated signal equal in power and bandwidth efficiency to ideal SSB modulation, without the stringent filtering requirement. With further processing the signal can be converted to a SSB signal with significant savings in the spectrum required. A popular technique employs two independent, equal-rate digital streams which are used to form two DSB-AM/SC signals. One signal is displaced 90° in phase from the other, and the two are added together. The result is known as *quadrature amplitude modulation* (QAM). If the two digital streams are encoded in three-level duobinary (+1, 0, −1), the result is known as *quadrature partial response modulation* (QPR). If the two digital streams are encoded in L levels, the result is a signal with L^2 states known as L^2 QAM (i.e., 8 levels produces 64 QAM) (A digital microwave radio employing 16 QAM is described in Section 4.1.1.1.2.)

A simple FM technique uses the binary signal to switch between two frequencies. This is known as *frequency-shift keying* (FSK). In its simplest form it is characterized by abrupt phase changes at the transition between one state and the other which distort the spectral energy and reduce spectral efficiency. A better technique is known as *continuous-phase FSK* (CPFSK), in which the phase transitions are smoothed by looking ahead to the next bit interval and anticipating the change to be made. Even better performance is achieved by limiting the frequency deviation to twice the data rate. This technique is known as *minimum-shift keying* (MSK).

Simple *phase-shift keying* (PSK) uses the binary signal to alternate the phase of a carrier between 0° and 180°. A variation, known as *differential PSK* (DPSK), compares the phase of the previous bit with the present bit to determine if a transition has occurred. Another technique encodes two bits at a time into one of four phase values spaced 90° apart. This is known as *quaternary PSK* (QPSK).

The purpose of these modulation schemes is to encode as much digital information (bits/second) into a given bandwidth as possible. Performance is often measured in terms of bits/second/hertz (b/s/Hz). Care must be taken in comparing modulation techniques on this basis. Actual performance depends greatly on the signal-to-noise (S/N) ratio, the bit-error-rate (BER, that is, the average number of bits which are received correctly for every bit which is

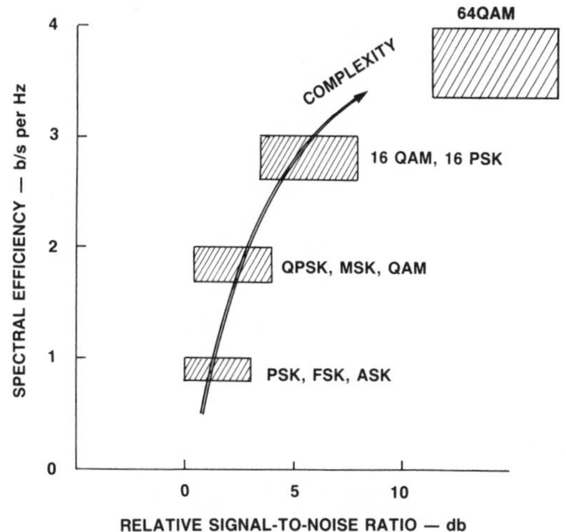

Figure 3.9. Performance of digital radio modulation techniques as a function of complexity. The location of the domains is determined by bit-error rate. Higher spectral efficiencies can be achieved at lower BER performance.

received incorrectly), and the degrading effect of a limited bandwidth transmit filter (the use of which is essential to limit radiation to the allocated bandwidth). The ranges of performance of four levels of complexity of modulation techniques are shown in Figure 3.9. Reliable efficiencies of around 1 b/s/Hz are achieved with simple phase, frequency, and amplitude shift keying. Two bits per second per hertz can be obtained through the use of 4-phase PSK (QPSK), MSK, and QAM. Three bits per second per hertz requires more complex schemes such as 16 phase PSK, and 16 state (2 streams with 4 levels) QAM. Four bits per second per hertz is achieved with 64 QAM. As the complexity of the modulation technique increases, a better signal-to-noise ratio is required to achieve the same BER.[(4)]

3.3.5. Spread-Spectrum Signals

Spread-spectrum signals occupy a bandwidth larger than the minimum bandwidth needed to contain the information they carry. Bandwidth spreading is accomplished through the use of an independent code which is impressed on the information stream to produce a composite signal which has a higher (usually, very much higher) bit rate than the original signal. This is often accomplished by *direct sequence* modulation in which each information bit is divided into smaller increments of time (called *chips*). PSK or FSK are employed to encode the chips so as to produce a signal containing many components of relatively

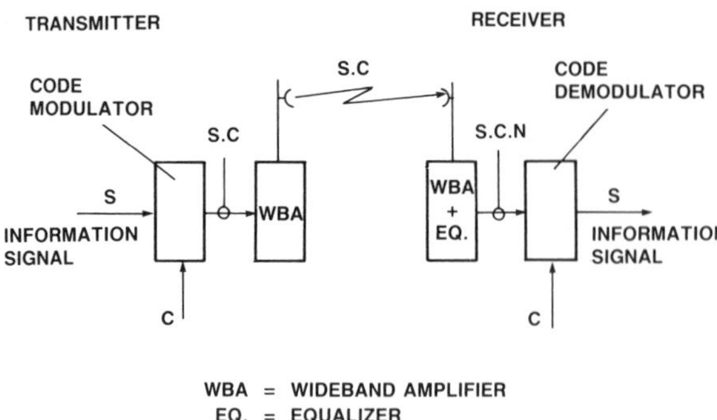

Figure 3.10. Principle of spread-spectrum telecommunication. Even though the signal (S) is corrupted by noise (N), use of the spreading code (C) in demodulation will produce an output signal virtually noise free.

low energy distribution over a wide frequency band.[5] The principle of spread-spectrum telecommunication is shown in Figure 3.10. The information signal (S) is modulated by code (C) to become the wideband signal (S.C) which is amplified and transmitted over a radio link. The signal arriving at the receiver is corrupted by noise due to environmental conditions and narrow-band interferors. It is amplified, equalization is applied to restore the signal-to-noise/frequency relationship to values commensurate with those existing at the transmitter, and demodulated with the aid of C to produce the information signal S. For demodulation to occur the receiver must *know* the spreading code employed at the transmitter. To a receiver which does not know the chip code, the signal appears noiselike.

Spread-spectrum techniques can be applied to short-range, lower-power transmission systems (such as cordless telephones, remote control of domestic appliances, cellular mobile telephone, etc.). By controlling the number of transmitters so that the total power spectral density of all simultaneous users does not exceed acceptable ambient radiated noise levels, these systems can be operated in frequency bands already allocated to other services. Further, by providing each transmitter with a code which is different from others operating in the same band, multiple simultaneous users can occupy the channel without mutual interference. This mode of operation is known as *code-division multiple access* (CDMA).

3.4. Television/Video Signals

Television signals are generated by imaging a scene on the optically active target of a camera tube while an electron beam scans it horizontally from top to bottom. In this way the picture is converted to a signal containing segments representing horizontal *lines* whose intensity is modulated in accordance with the details of the scene. When the beam reaches the bottom of the target it returns to the top of the picture to start again. In the time (known as *vertical interval*) required for this action *(vertical retrace)* a small number of unmodulated lines are produced.

In order to reduce the bandwidth required, yet provide a relatively flicker-free picture, the scene is scanned in halves: thus, all odd-numbered lines are traced, then all even-numbered lines are traced. These two frames (odd and even) comprise a complete picture. In those countries which employ a basic power-line frequency of 60 Hz, the picture usually contains 525 lines (495 lines picture, 30 lines vertical retrace) and a complete picture is scanned 30 times a second. In those countries with 50 Hz power, the picture usually contains 625 lines (575 lines pictures, 50 lines vertical retrace) and a complete picture is scanned 25 times a second.

3.4.1. NTSC Color Signal

Around the world, three color systems are in use: NTSC (in the U.S.A. and Japan), PAL (in most of Europe), and SECAM (in France and Russia). Only

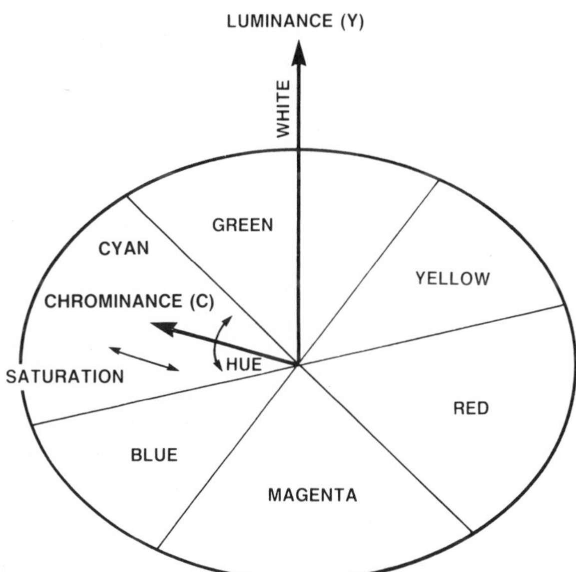

Figure 3.11. An NTSC color television signal has two components. Luminance is the measure of brightness or brilliance and is the part of the signal used by black-and-white receivers. In the chrominance signal, phase determines the absolute color, and amplitude determines the vividness of that color.

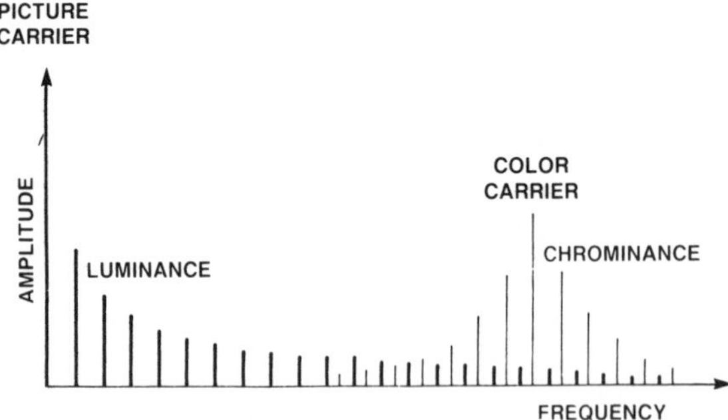

Figure 3.12. Representation of the NTSC color signal showing color carrier and chrominance signal placed between the harmonics of the luminance signal.

the NTSC (National Television Standards Committee) system is discussed here. The signal, which can be used by both black-and-white and color receivers, contains two components: *luminance,* which is proportional to the brightness or brilliance of each element of the scene, and *chrominance,* which is proportional to both the hue and the saturation of each element. A representation of these components is given in Figure 3.11. Luminance measures the picture intensity. It is that part of the signal which is used by black-and-white receivers. The phase of the chrominance signal represents the color or hue, and its amplitude represents the boldness of the color (i.e., whether pastel or saturated). Chrominance information is contained in an amplitude- and phase-modulated 3.58-MHz signal

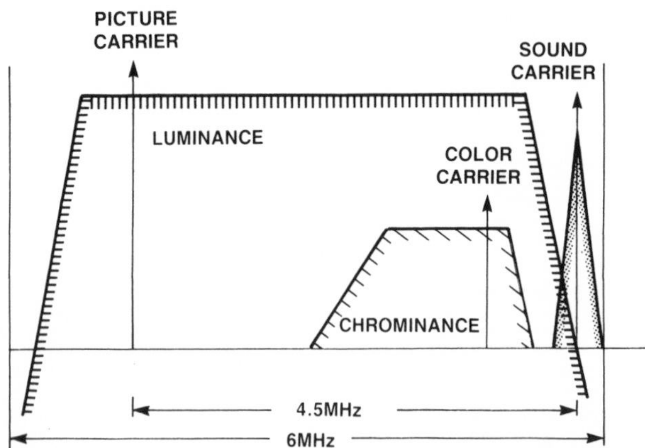

Figure 3.13. A broadcast color television signal comprises luminance, chrominance, and sound signals. Both luminance and chrominance employ VSB-AM. The sound signal is FM.

which is carried on the luminance signal. A short burst of the unmodulated chrominance signal (color carrier) is inserted at the beginning of each line to provide a stable phase reference. In the frequency domain, the luminance and chrominance signals fit within each other. The frequency components of the chrominance signal are interleaved with the frequency components of the luminance signal, as illustrated in Figure 3.12. For broadcasting, the composite signal is modulated on an appropriate carrier using VSB-AM and an FM sound channel is added. The make-up of a complete broadcast television channel is shown in Figure 3.13. It requires a channel bandwidth of 6 MHz. Proposals for higher-definition television are discussed in Section 4.4.4.1.

3.4.2. Digital Television

A color television signal can be digitized using the sampling rule stated in Section 3.1.1. Using 6 MHz as the maximum frequency to be reconstructed (analogous to 4 kHz for voice) and quantizing the amplitude into 128 or 256 levels (7 or 8 bits) gives a bit rate of 84 or 96 Mb/s. Without jeopardizing picture quality, it can be reduced by recognizing that the video information is contained in 4.5 MHz, and the sound can be transmitted on a separate channel. Thus, the video bit rate can be reduced to 63 or 72 Mb/s. Further, the significant high

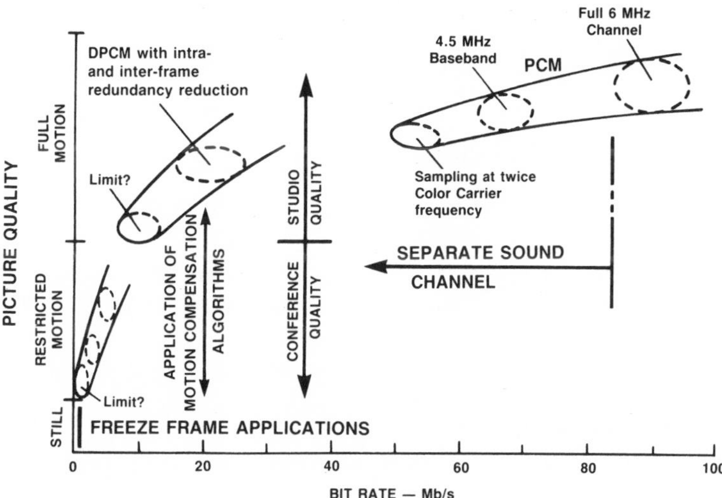

Figure 3.14. Summary of digital television capabilities as a function of bit rate. Full-motion video can be provided in the range of 20 to 100 Mb/s. The lower limit may be dropped to around 10 Mb/s through application of motion compensation processing. Restricted motion video suited to videoconferencing can be provided at rates of 1.54 Mb/s and less. Freeze frame applications use bit rates as low as 1200 b/s.

frequency is 3.58 MHz, the frequency of the color carrier. By sampling at twice this frequency, the lower sideband of the chrominance signal will be preserved, and only the very highest frequency components of the luminance signal will be destroyed. So, for a small degradation of picture quality, the bit rate can be reduced to approximately 50 to 57 Mb/s.

By using the difference between samples (differential PCM), reducing redundancy on a line-by-line basis (run length coding), and reducing redundancy on a frame-by-frame basis, the bit rate can be lowered to the range of 20 to 30 Mb/s. Modern equipment is available which will transmit network quality, full-motion video at these rates. For situations in which motion is restricted, such as videoconferencing, the bit rate can be reduced to 6.3 Mb/s and lower. With motion compensation, the bit rate for network quality, full-motion video may be reduced to around 10 Mb/s, and for restricted motion videoconferencing to less than 1.5 Mb/s. Motion compensation is achieved by identifying the area of the picture which is in motion and deriving a motion vector from frame to frame. Transmitting the combination of area and vector allows reconstruction without a bit-by-bit message. Systems are becoming available which produce acceptable conferencing pictures at 56 kb/s. Figure 3.14 illustrates the relation between bit rates and video quality for these techniques.

3.5. Transmission

Telecommunication involves transmitting diverse signals over a distance by various means. A distinction is made between broadcast and guided media, which behave somewhat differently. All media carry multiple channels in order to reduce the cost of transmission.

3.5.1. Transmission Means

In free space, messages are transmitted by radiating modulated signals whose energy is all contained within the bandwidth of the channel (so as not to interfere with other users). Messages are also transmitted over means which guide or contain the energy, such as wires, coaxial cables, waveguides, and optical fibers.

Broadcast and guided transmission paths exhibit an important difference in regard to the power required to communicate over long distances. For broadcast signals, the energy flow between transmitting and receiving antennas is governed by the principles of electromagnetic wave propagation first enunciated by James Clerk Maxwell (Maxwell equations). For ideal (isotropic) antennas

$$P_R = P_T(\lambda/4\pi d)^2 \tag{3.4}$$

where, P_R is the received power; P_T is the transmitted power; λ is the wavelength, i.e., the distance traveled by a point on the wave during one cycle; and d is the distance between the antennas. Doubling the distance over which a given power level is to be transmitted requires quadrupling the power radiated. Thus, if a broadcast transmitter of 1000 W (60 dBm; 0 dBm = 1 milliwatt) carries a given distance, a transmitter of 4000 W (66 dBm) will carry twice as far. The transmission loss L is given by

$$L = 10 \log_{10} (P_T/P_R) \text{ dB} \quad (3.5)$$

For signals which are carried over means which guide the energy, the attenuation of each section acts on a signal already diminished by the attenuation of previous sections. For each section, $P_i = aP_o$, where P_o is the power out and P_i is the power in: hence, for a link of n sections,

$$P_T = a^n P_R \quad (3.6)$$

and

$$L = n \cdot 10 \log_{10} a \text{ dB} \quad (3.7)$$

Thus, in a fiber optics transmission system with a total attenuation of 1 dB/km (for instance), the loss along a 30-km path will be 30 dB. A path twice as long will result in a loss of 60 dB. If the original path (30 km) is adequately served by a laser diode emitting 1 mW (0 dBm), doubling the length will require an input power of 1 W (30 dBm)—a level which is not attainable with today's devices. Quadrupling the power of the laser (4 mW, 6 dBm) will increase the path length to only 36 km. Because of the different ways in which they respond to increased input power, broadcast systems employ relatively high powers and guided systems employ low powers.

The propagation of electrical signals along wire cables was studied by many early telegraphers. In 1881, Oliver Heaviside correctly stated the relationship between cable resistance (R), inductance (L), and capacitance (C) in what are known as the *telegrapher's equations*. He demonstrated that signals travel at a constant speed dependent on the value of the line parameters (R, L, and C/unit length) and that, with a certain relationship between them, a propagation speed and attenuation is achieved which is independent of frequency, thereby producing an undistorted signal. Achieving this state with today's wire pair cables requires adding inductance to the cables. Known as *loading,* this technique made it possible to operate wire pairs over longer distances (see Section 4.2.1). By 1899, practical loading coils had been developed for overland use, and continuously loaded submarine cable was perfected in 1902.

3.5.2. Multiplexing and Concentration

The cost of transmission facilities increases with the distance to be covered. For this reason, all but the shortest connections carry as many channels as possible so as to reduce the cost per channel. Two basic techniques are used: frequency-division multiplex (FDM) and time-division multiplex (TDM). In FDM, a set of channels equally spaced in frequency are used to create a composite signal. Figure 3.15A illustrates how 12 message channels can be *multiplexed* by steps to become part of a signal containing 600 or more channels. In TDM, a specific time interval (frame time) is divided into equal segments which are assigned to

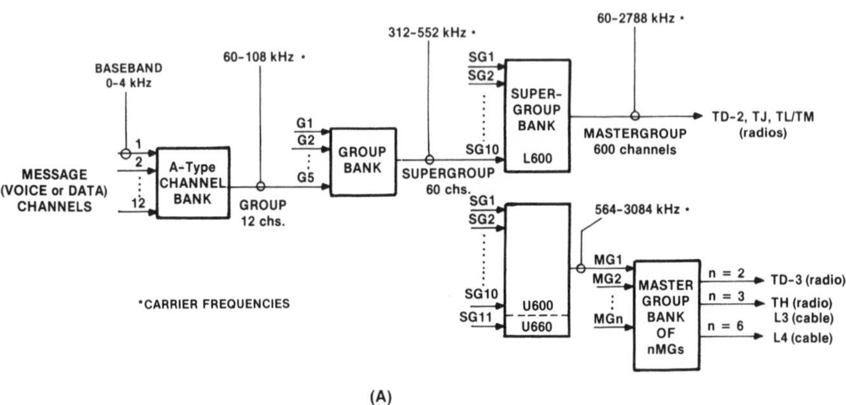

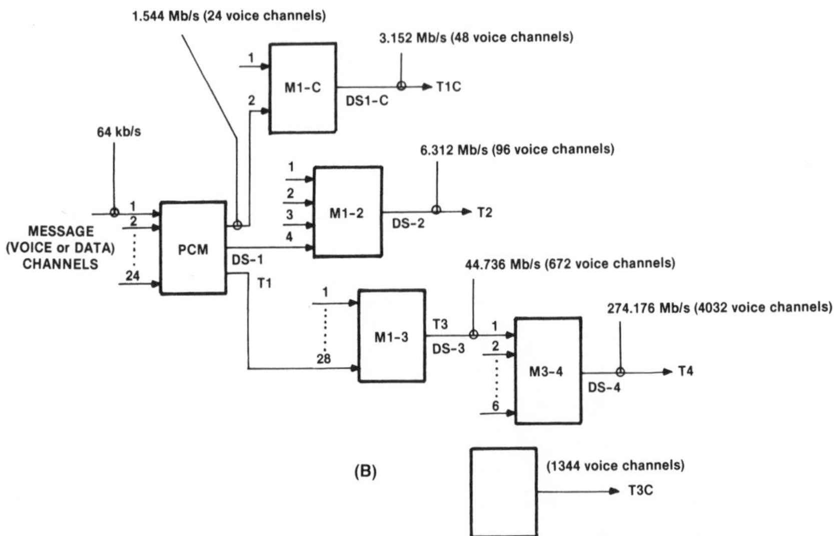

Figure 3.15. Multiplexing of message channels. In (A), twelve 4-kHz channels are frequency-division multiplexed to become part of a 600-channel master group. In (B), twenty-four 64-kb/s channels are time-division multiplexed to become part of 4032 channels in a T4 system.

each channel. The individual digital signals are combined in successively higher speed bit streams. Thus, 24 voice channels (64 kb/s each) are multiplexed to produce a bit stream of 1.544 Mb/s (T1). The process requires the collection, storage, and manipulation of each signal and delays each frame by 125 microseconds, the time necessary to assemble 8 bits. Figure 3.15B illustrates how 24 channels can be multiplexed by steps to become part of a 274 Mb/s stream containing 4032 channels.

A third technique is known as statistical multiplexing (SM). Each user is assigned a channel when information is to be transmitted. When no information is ready, the channels are used by others. Invented for, and used on, submarine telephone cables, the technique is now used on terrestrial and space links. By taking advantage of the silent periods in speech, a doubling of the capacity of even moderate size trunks can be obtained. For data terminals operating with packets, greater efficiencies can be realized, and packet operation incorporates this effect. (SM is discussed further in Sections 4.1.1.5.2 and 4.1.1.6.)

In contrast to multiplexing, which ensures all input channels are provided frequencies or timeslots in the output channel, *concentration* makes a fixed number of output channels available to a larger input community. If all require channels simultaneously, some cannot be served. Concentration is used in situations where the average channel is little used and the probability is very small that all input channels will require service at the same time. The common (shared) equipment is supplied to a level which will maintain a prescribed grade of service. For instance, in a residential environment, not all telephones are used at the same time, so it would be extravagant to equip the end office to provide enough equipment to service every customer simultaneously. In a business environment, the number of lines available from PBX to end office is substantially less than the number of telephones served by the PBX. The implementation of concentration is guided by traffic analysis and service levels, subjects discussed in Section 3.7.

3.5.3. Electromagnetic Spectrum

Figure 3.16 shows the *electromagnetic spectrum,* and the major applications for which frequencies have been assigned. In theory, electromagnetic waves can be generated at any frequency. In practice, a lower limit is established by the impracticability of constructing large enough antennas. At a frequency of 60 Hz, for instance, the *wavelength* (wavelength = c/f, where c is velocity of light and f is the frequency of the wave) is approximately 5000 km (3125 miles)—a distance greater than the distance from Boston to San Francisco. It would be extremely difficult to launch such a wave efficiently.

The earth is surrounded by an ionized layer known as the *ionosphere* which is produced by the impact of the sun's rays on the gas molecules of the upper atmosphere. Up to 30–50 MHz, electromagnetic energy hugs the surface of the

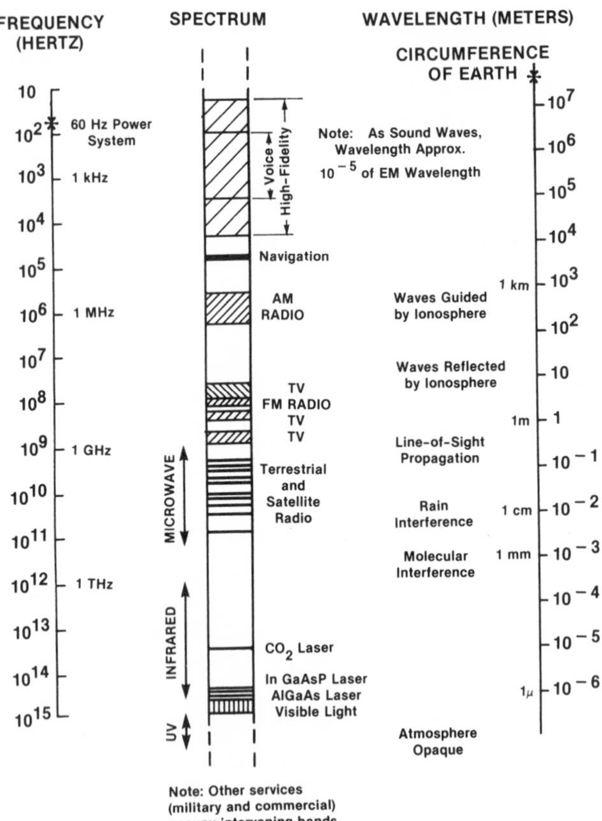

Figure 3.16. The electromagnetic spectrum showing the relationship between frequency, wavelength, applications, and propagation effects.

earth or is reflected by the ionosphere (or both), making long-distance broadcasting possible. Above 50 MHz, the energy is neither reflected by the ionosphere nor conducted along the earth, but propagates in straight lines, so that reception depends on the transmitter being in the *line of sight* of the receiver. At microwave frequencies, that is frequencies above 1 GHz, antennas are easily constructed to radiate in narrow beams. They are used for satellite and point-to-point terrestrial communications. At frequencies above 10 GHz, intense rainfall interferes with propagation, and above 20 GHz, water and oxygen molecules present in the atmosphere are significant energy absorbers, limiting communications to *windows* between these absorption peaks.

Bands of frequencies are allocated to different services by government bodies on the basis of agreements reached at World Administrative Radio Conferences (WARCs) established to organize the use of the spectrum and reduce interference between applications to a minimum. Users of the spectrum are licensed and regulated by government. They are assigned a separate *channel* so as not to interfere with other users. Channel width is established on the basis of acceptable, not perfect, performance. In the United States, AM radio stations are allocated 10 kHz channels over which they broadcast DSB signals of less than 5 kHz baseband, FM radio stations are allotted 75 kHz channels over which they broadcast monaural signals of 15 kHz baseband or multiplexed signals which are used for stereo and other services, and television stations are allocated 6 MHz channels over which they broadcast video signals of 4.5 MHz baseband. That AM radio and television stations must broadcast less bandwidth than full fidelity requires is a reflection of the finite spectrum available and the intense competition for its use.

Optical frequencies, which lie beyond 1 THz, are divided into infrared, visible, and ultraviolet bands, and are normally specified by wavelength, not frequency. Thus, infrared extends from 0.8 to 100 μm (1 μm = 1 millionth of a meter = 0.00004 in.), visible light occupies wavelengths from 0.35 to 0.8 μm; and ultraviolet has wavelengths below 0.35 μm. A carbon dioxide gas laser used for welding and cutting produces optical energy at a wavelength of 10 μm, and solid-state lasers made from AlGaAs and InGaAsP materials produce energy at 0.85, 1.3, and 1.55 μm for use in optical fiber transmission systems (see Section 3.12.4).

3.5.3.1. Frequency Reuse

At microwave frequencies the radiated energy can be shaped into pencil beams. For terrestrial radiolinks, energy at the assigned frequency is directed to the receiving antenna. A few degrees off the line of sight, and the radiation is undetectable. This means that a second transmitter–receiver pair can use the same frequency if their line of sight is separated adequately from the first. *Frequency reuse* is an important technique which has been essential to the growth and development of microwave radio communications.

From space, communication satellites can broadcast to a large fraction of the earth's surface, or concentrate energy on specific areas. A spot beam which may be a hundred miles or so in diameter on the surface can use all of the frequencies assigned to the satellite, and other spot beams directed to areas perhaps 1000 miles away can reuse the same frequencies. Earth stations employ relatively large antennas which focus their radiation into beams of well under 1° so that satellites only 2° apart may use the same frequencies.

Frequency reuse is also employed with terrestrial broadcast systems. By limiting the power radiated from an omnidirectional antenna, a circular cell of five or ten miles in diameter can be established in which communication between mobile receivers and the fixed antenna can be supported. Depending on the frequencies used, some 10 to 20 miles away a second antenna using the same frequencies as the first can be installed to establish another cell. A third cell is located between them. It employs different frequencies. Provided their receivers can be retuned to the frequencies in this third cell, mobiles can maintain communication while passing between the first two. (Mobile telecommunication is discussed in Section 4.2.2.)

Techniques other than physical separation can be used to achieve frequency reuse. For instance, the same frequencies can be employed in beams with *horizontal* and *vertical* polarizations. (The electromagnetic energy is launched with the electric vector horizontal for one beam and vertical for the other.) Using *polarization-sensitive* antennas, the two waves can be separated. Another technique employs time division. For a short time period (time slot) the entire frequency band is allocated to one transmitter–receiver pair which exchange a burst of information. The next time slot is assigned to another pair, and so on. After all active links have been served, the time slot is allocated again to the first link. Called *time-division multiple access* (TDMA) this technique is discussed further in Section 4.1.1.4.1.

3.5.3.2. Diversity Reception

Reliability is an important requirement for transmission links employed in public and private networks. Besides equipment reliability, which is of concern in all types of equipment, over-the-air transmission facilities must contend with the uncertainty of propagation through the atmosphere. Ionospheric shifts, sunspots and electromagnetic phenomena, rainstorms, fog, dust, and other obstructions, produce fading and polarization shifts which can reduce link performance to unacceptable levels. At installations where these effects occur frequently enough to be of concern, *diversity* techniques can be employed to restore operation to required levels. Thus, the link may employ two separate frequencies chosen so that the probability of poor performance on both links simultaneously is significantly less than the probability of poor performance on one or the other. *Frequency diversity* can be very effective; however, since it requires two separate frequencies, it can only be used on the most important links. Another technique uses *space diversity*. For satellite communications, receiving earth stations may be separated by several miles so that poor atmospheric conditions at one site are unlikely to spill over to the other site at the same time. For terrestrial radio links, two microwave receiving antennas may be mounted on the same tower and separated by as little as 30 ft to overcome atmospheric effects along the line of sight.

3.6. Digital Switching

Switches selectively establish and release connections between transmission means. They make information flow possible between any two points in a network, allow the reuse of facilities (to carry other messages), and serve to con-

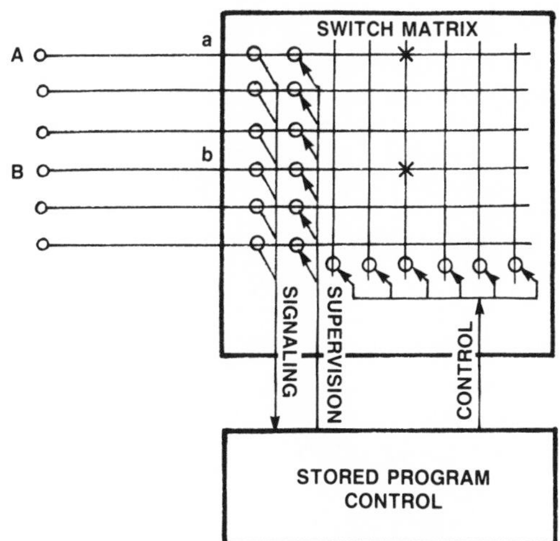

Figure 3.17. Principle of a stored program controlled (SPC) switch. Connection between users A and B is made by the control unit software which configures the switch matrix to provide a circuit between port *a* and port *b* upon receipt of signaling information from the caller (A). The switch matrix may employ space-division or time-division techniques.

centrate traffic so as to make efficient use of transmission paths. Two modes are in use: circuit switching and packet switching. The former is typical of the telephone network; the latter is used in data networks, and will be described later in Section 4.7.4.1.

Modern circuit switches are computer controlled. Often, they are referred to as *stored program controlled* (SPC) machines, to differentiate them from older, electromechanical switches, such as step-by-step (S × S, also known as Strowger) and crossbar (Xbar). All-digital, time-division switches are rapidly replacing these electromechanical switches (as well as displacing earlier SPC switches which use analog techniques) because of the space and power savings, the reliability improvement which is achieved through the use of solid state integrated circuits, and the automated system management which is possible with an all-digital system. The principle of an SPC switch is shown in Figure 3.17. Connection between two users, A and B, is completed under software control upon receipt of signaling information from the caller and status information from the called party's terminal. During the call setup phase, the SPC software assigns switch resources and provides call progress signals (such as dial tone, ring back, station busy, etc.). The switch matrix may employ *space-division* or *time-division* techniques, or a combination of the two, to maintain a circuit between A and B for the duration of the call.

A space-division matrix provides a physical path between the parties by means of electromechanical or electronic cross-points. Electromechanical cross-points (such as reed relays) pass signals in both directions so that A and B can communicate over a two-wire circuit. Electronic cross-points are unidirectional devices and two cross-points must be employed, one for the *go* path and the other for the *return* path. A space-division matrix is relatively easy to implement,

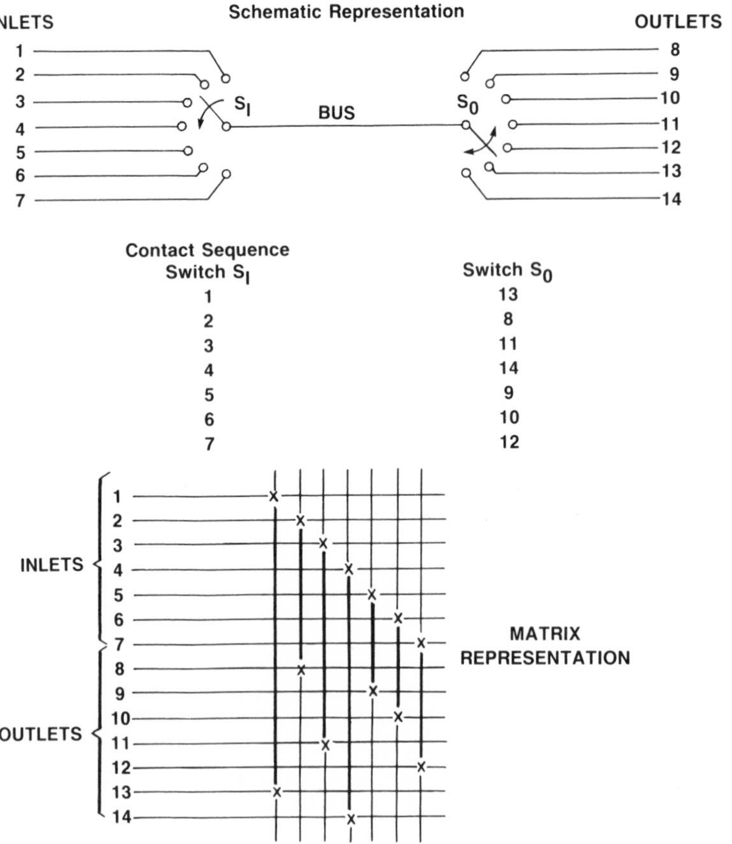

Figure 3.18. The principle of time-division switching. For a short time inlets 1 through 7 are connected in sequence to outlets 13, 8, 11, 14, 9, 10, and 12, respectively. Rapid repetition allows information to flow between each inlet and outlet.

and, unlike time division, the technique does not place an intrinsic limitation on the bandwidth of the signal which can be switched.

In a time-division matrix, a time slot is associated with each connection. During this period, information is passed on the *go* connection and on a separate *return* connection. In principle, for a short time, inlet and outlet are connected by high-speed switches. Figure 3.18 illustrates the operation. Inlets 1 . . . 7 are connected in sequence through switch S_I to switch S_O. Switch S_O moves from outlet to outlet in such a way as to connect the proper outlet to each inlet. Thus, for the example given: when S_I is on 1, S_O is on 13; when S_I is on 2, S_O is on 8; etc. Rapid repetition of the sequence allows information to flow from inlets to outlets. Each of the seven inlet signals occupies the bus sequentially, and S_O

connects them to the appropriate outlet. Changing the sequence in which the outlet switch switches alters the connection pattern. For analog signals, time-division switching results in strings of pulses of varying amplitudes from which the outlet signal is reconstructed. If the switching process is not to distort the inlet information, it follows that switching must occur at least twice as rapidly as the highest frequency the input information contains. Put another way, the speed at which a time-division switch operates limits the bandwidth of the information it can switch.

For digital voice signals encoded in 64 kb/s PCM, 8-bit words must be transferred 8,000 times a second. Moreover, transfer should only take place when each word is complete, and the transfer of complete words from different voice signals to a common multiplexed stream must be possible. These requirements make it necessary to employ a common synchronous action within the switch, and to extend the synchrony to all other directly connected digital facilities supported by the switch.

In the schematic representation of Figure 3.18, the number of digital voice channels which can be switched by S_I or S_O is limited by the bit rate supported by the bus. For a practical switch, we can imagine several parallel sections, each of which contains two switches connected by a bus. They interconnect specific groups of inlets and outlets. To make any outlet available to any inlet requires the use of a switch to interchange traffic from any bus, or highway, to another. Known as a highway interchange (HI) switch, it can be implemented as a digital space (S) switch as shown in Figure 3.19. Incoming multiplexed highways are

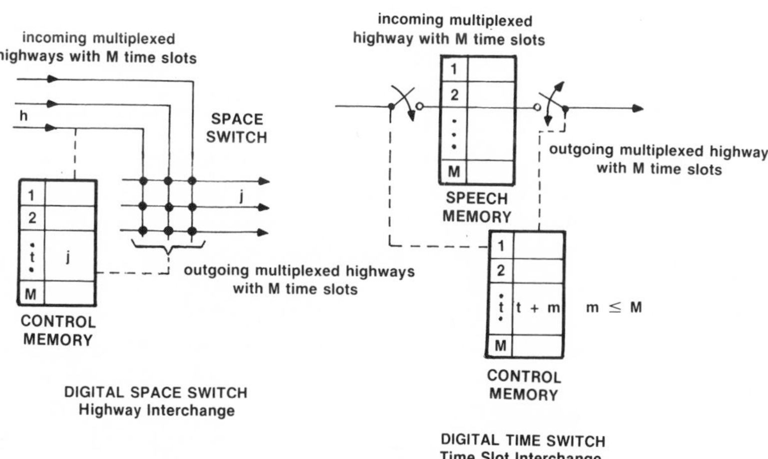

Figure 3.19. Principle of highway interchange and time-slot interchange in time-division space and time switching.

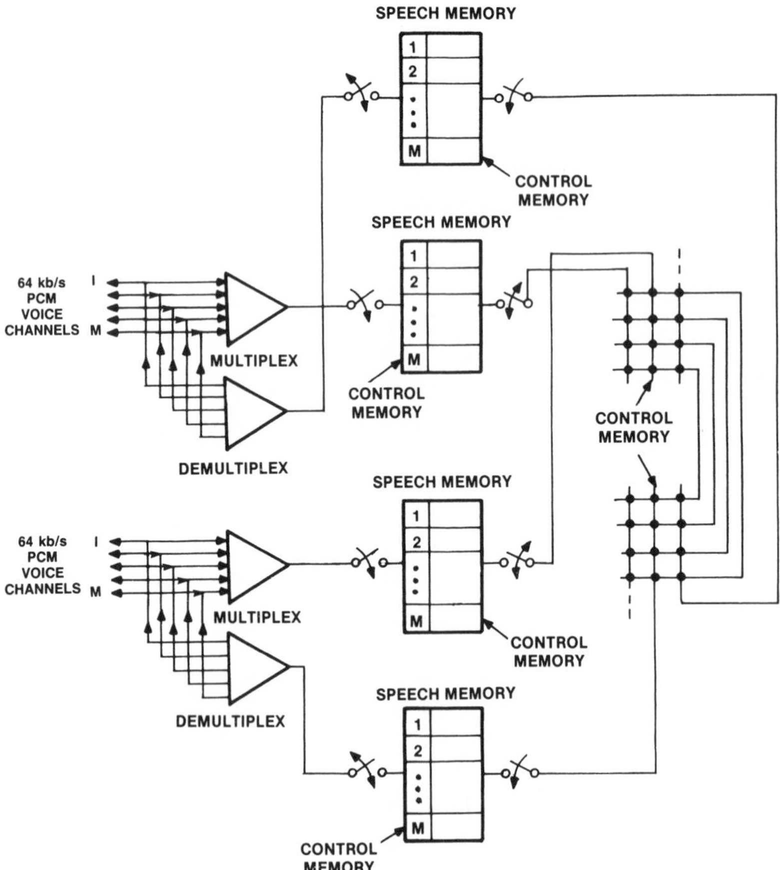

Figure 3.20. Principle of digital switch employing time/space/space/time switching.

connected to outgoing highways by means of a matrix of electronic gates which are activated by information written in the control memory. In this way a word in timeslot t on incoming highway h is transferred to timeslot t on outgoing highway j.

Our simple model does not allow for time slot interchange, that is, the storage of incoming information so that it can be inserted in a later time slot in the output stream. Time slot interchangers (TSIs) can be implemented as digital time (T) switches as shown in Figure 3.19. Each incoming word is stored in sequence in memory (speech memory) and read out in a different sequence under the direction of information contained in the control memory. In this way a word in timeslot t contained in a frame of M timeslots at the input can be transferred to timeslot $t + m$ (where $m \leq M$) at the output. (This manipulation results in a delay of as much as one frame period.)

Put together in various arrangements, these space and time switches can form digital switches of any size. The principle of a digital voice switch which employs time, space, space, and time (TSST) switching units is illustrated in Figure 3.20. For small switches, the order of the switching stages has no major impact on cost. Switches employing TST or STS architectures are equally costly. For larger switches, a TSSST architecture is around 25% less costly than STSTS. For very large switches a TSS–SST arrangement is significantly less costly then STS–STS.[6] The elements of switching software are discussed in Section 3.8. Examples of modern switches are given in Chapter 4.

B. CONTROL

In this section, we discuss traffic theory, real-time software methodology, artificial intelligence applications, signaling, and data network protocols. They are all important in understanding the complexities of the control and management of telecommunication facilities.

3.7. Traffic Theory

Telecommunication facilities are connected together in such a way that many customers can make concurrent use of them to satisfy their individual communication requirements. Because only a fraction of all customers originate calls at any one time, the amount of equipment required to carry the traffic is much less than what would be needed to carry traffic from everyone, simultaneously, and modern telecommunication systems are built to share facilities. This resource sharing implies that there will be times when two or more parties contend for the same facility—and some requests for connections cannot be processed immediately. This situation occurs when the arrival rate of requests for service increases above the rate which can be served, or the service facilities available decrease below the number needed to serve the requests. What happens depends on whether the unserved requests wait their turn, or are abandoned. Familiar examples are provided by *dial tone* and *busy tone* in telephone installations. On picking up the handset, the caller expects to receive dial tone immediately. Should too many persons attempt to place calls at the same time, the capacity of the common equipment is exceeded and it may be several seconds before the caller receives dial tone. In this case, the request for service was held until equipment became available. If the wait becomes many seconds, the caller may abandon the call and start again, or the equipment may place an equipment busy tone on the line requiring the caller to hang up and start again. The results are reflected in the *grade of service* provided the customer and depend on statistical quantities such as request arrival rate, holding time, and traffic intensity.

3.7.1. Request Arrival Rate

Dividing the number of requests for service received during a measured time interval by the interval gives the average request arrival rate. In practice, the rate varies significantly with time of day. What is important is the rate associated with peak demand. Denoted by λ, it is usually expressed as the average number of requests in one second of a typical peak hour. Multiplied by 3600 (the number of seconds in an hour) it is the number of busy-hour attempts (BHAs).

3.7.2. Holding Time

Once a request for service has been accepted and a communication path established, message transfer can proceed. The length of time the circuit remains connected is the holding time. It may vary from a few seconds for a short data message (such as credit verification) to many hours for a teleconference. For telephone conversations, approximately half the calls last longer than one minute, one-third last longer than two minutes and one-sixth last more than three minutes. Holding time is denoted by h (sometimes CH or TH) and is expressed in seconds.

3.7.3. Traffic Intensity

The amount of traffic due to λ requests per second over a period of time T seconds, and an average holding time of h seconds is $T\lambda h$. This is often expressed in units of *hundred call seconds* (ccs). 1 ccs is the traffic due to one call of 100 seconds duration, 100 calls of 1 second duration, or any combination in between. Traffic is also expressed in *erlangs*—a unit named after A.K. Erlang, a Danish mathematician who pioneered much early telephone traffic theory. The number of erlangs is the ratio of the time during which a circuit is occupied to the time for which it is available for occupancy. Traffic which occupies a circuit for one hour (a busy hour) is equal to 1 erlang; thus, 1 erlang is equal to 36 ccs/hr.

3.7.4. Grade of Service

Should the number of service requests exceed the maximum number which any part of a telecommunication system can handle, they will be queued or rejected. For most communication systems, particularly public systems, there are periods during the day when service requests are at their peak. To engineer a system to accommodate these levels without degradation requires a significant investment in equipment. To the extent that this investment is not made, service will deteriorate during periods of peak use.

Grade of service is a numerical quantity which describes the level of service provided. It includes all of the factors which delay or prevent a message being sent. In a telephone system, dial tone response time and call setup time are two parameters which delay the process. Unavailability of common equipment blocks the process. In other systems similar factors affect completion. Determination of grade of service requires the collection of a statistically significant amount of performance data. Expressed as *1* or *100%*, a perfect grade of service is rarely attained in large systems. Satisfactory performance is usually denoted by values in the high nineties.

Three basic expressions (Erlang B, Erlang C and Poisson) exist for calculating grade of service, that is, the probability of requests for service being blocked, as a function of the number of servers m and offered load λ/μ, where λ is the request arrival rate, already specified, and μ is the request departure rate, per server. The *Erlang B* formula assumes that blocked requests are simply lost to the system: the *Erlang C* formula assumes that blocked requests are delayed but eventually served: the *Poisson* formula lies between these cases. For all three models, tables of numerical values are readily available for use by those who must project system traffic performance. In addition, other formulas exist, including combinations and modifications of these basic relationships.[7] All are dependent on the assumptions employed to define the statistics of request arrival and what the initiator of a blocked request is likely to do. All assume a very large number of requests are made so that the statistical expressions are valid. None predicts what happens to an individual request for service, and actual experience over a short time interval will most likely not validate the calculated results. Further, specific applications may generate request arrival statistics which do not match the model intrinsic to these expressions. Despite all of these limitations, designers depend heavily on these formulas to estimate the number of servers required throughout their systems.

3.7.5. Quality of Service

Quality of service is a concept which deals with the overall quality of the connection once it has been made. It includes the effect of all of the factors which distort, destroy, or introduce error into the message. Thus, for a voice message, frequency distortion may make the message difficult to comprehend, severe echos may destroy the intelligibility of the message, and electromagnetic interference may introduce extraneous noises. Quality of service under these circumstances is a subjective judgment of the user which can be elicited by a trained observer using a carefully designed set of questions. For a data call, quality of service is largely determined by the bit-error rate. *Quality of service* should not be confused with *grade of service,* which reports on the result of contention for shared resources and defines the probability of being connected.

3.8. Software

In common with other electronics based activities, telecommunication is depending to an increasing extent on the use of computers to perform control and processing functions which affect the transportation of messages, and to provide flexible features for the user. Much of this capability resides in *software*—the set of instructions (programs) which drive the hardware to perform as required. For many contemporary systems, the cost of software development far exceeds the cost of hardware development. Because it is invisible, the designer must construct an intellectual model of the intended operation of a program. In doing this, the reality of a set of instructions which produces a sequence of operations in the computer in a deterministic way can be lost. To assist in this abstract activity, a set of principles has been formulated which introduce structure and order to the effort, and form the basis for software engineering.

3.8.1. Software Engineering

The discipline known as *software engineering* places limits on the analyst's ability to construct software according to individual preference. By imposing rules and restricting options, it makes large software development teams possible in which each member's contribution is understandable by all other members, and is integratable into a single product. Further, the structure and function of the programs can be understood by others (verifiability), modules can be replaced (replaceability), or new modules can be added (extendability) without disturbing the design, and modules can be used in other products (transportability). Verifiability produces confidence in the integrity of the design through review by peer groups. Replacement and extension allow functions to be eliminated, changed, or added without penalty. Transportability is a powerful concept—it holds out hope for reducing the cost of future systems. These benefits are achieved in structured programming through the application of three main concepts: top-down design, a limited number of control statements, and data abstraction.

3.8.1.1. Top-Down Design

Top-down design begins with a simple statement of the objective, which is decomposed methodically to greater and greater levels of detail until it has been divided into clearly distinct and relatively independent parts. After establishing top-level control and data definition, work proceeds downward through functional modules and data structures to single functions which are small enough to be programmed, tested, and integrated by individual analysts. By restricting data flow between levels to the functions and procedures offered by lower levels, a high degree of independence between the parts of the program can be achieved. Step-by-step refinement of the design permits continuous review and systematic identification of areas of difficulty—which can be resolved before coding—making testing relatively trouble free and uncomplicated.[8]

3.8.1.2. *Control Statements*

A program is a set of instructions which controls the functions performed by a computer. The simpler these control statements are, the easier it is to understand the sequence of activities. The fewer control statements employed, the easier it is to understand the program. The statements which have the simplest

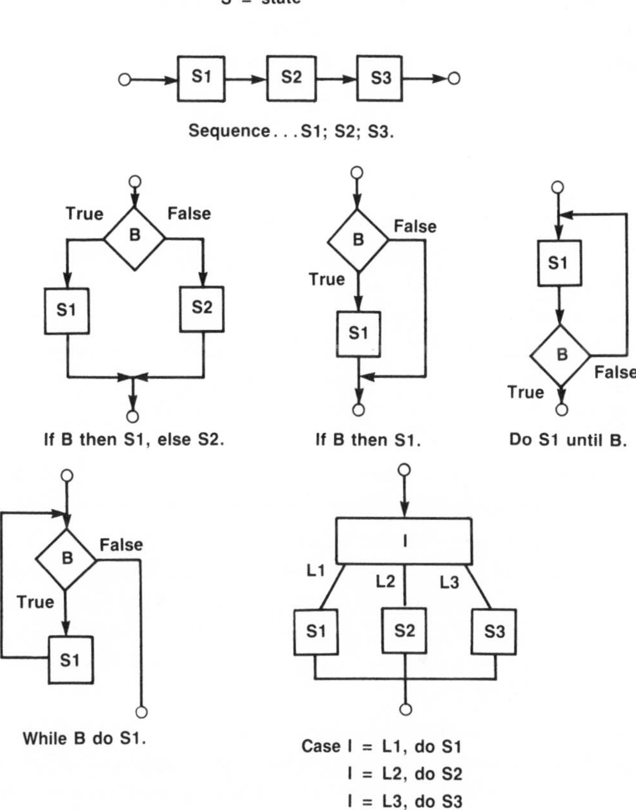

Figure 3.21. Basic statements for structured programming.

control flow are those which have a single entry and a single exit point. Although there is not total consensus on the essential minimum set of constructs, most practitioners use no more than the statements shown in Figure 3.21. More complex operations called *functions* or *procedures* can be constructed using these basic statements.

3.8.1.3. Data Abstraction

Most programs include information about objects or things which are declared as data, defined as to type (real, integer, etc.), and which are used by one or several modules as processing proceeds. An important concept in structured programming limits access to specific data files (structures) to as few modules as possible. Where this is impossible, *abstract data types* may be defined which consist of related collections of information applicable to one or more modules. They include limits on the range of data which is used and on the operations which may be performed using it. In this way data access is localized and controlled, and changes may be made in the basic data without concern for the many ways in which it is used in the program.

3.8.2. High-Level Languages

The languages used throughout the design cycle have a major effect on the integrity of the program product. In general, one language may be used for requirements specification and perhaps high-level design; a second language may be used for high-level design, detailed design, and coding; and a third language may be used for testing and man–machine interface implementation. However, the sets of statements which result from the use of these languages are unintelligible to the computer. Yet a fourth language is required, the language which is recognizable by the target machine. The first three languages are usually described as high-level languages, and the fourth is known as machine code. The specification language may be machine processable to check for integrity and completeness, but it is not a programming language, and cannot be used to produce computer instructions. The second and third languages can be used to define the operations which the machine will perform. They have the advantage that they are readily understandable by a trained programmer and each high-level statement creates several machine instructions. Using them, the programmer can solve the design problems at an abstract level rather than contend with the details of machine architecture inherent in machine code. Conversion of the high-level statements to machine code is performed by a *compiler*. An example of all three kinds of high-level languages is the set designed by CCITT (Comité Consultatif International Télégraphique et Téléphonique). They are SDL (Specification and Description Language), CHILL (CCITT High-Level Language), and MML (Man–Machine Language). The span of application of each is shown in Figure 3.22.

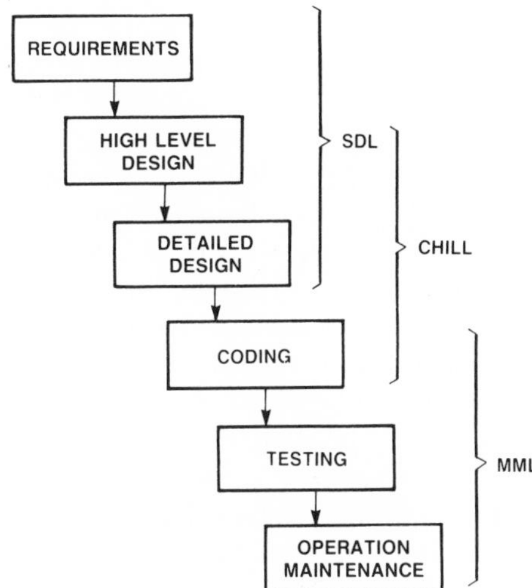

Figure 3.22. Span of application of CCITT high-level languages.

3.8.2.1. CCITT Specification and Description Language (SDL)

SDL describes telephony functions in language understandable by telephone-oriented persons. It specifies and describes the logic of functions and processes in a way which is independent of the implementation techniques that may be employed. SDL is documented in graphs containing unique symbols which represent subconcepts of a process (state, input, task, output, decision, and save) connected by directed flow lines. In order to make them more easily understood, these charts can be enriched by pictorial options which identify the equipment involved.

3.8.2.2. CCITT High-Level Language (CHILL)

A CHILL program consists of three parts: a description of data objects, a description of actions to be performed on these objects, and a description of the program structure. The language has been designed specifically for use in structured programs which control telecommunication switching facilities.

3.8.2.3. CCITT Man–Machine Language (MML)

MML is concerned with the interface between a person and an electronic switching system. It describes inputs to the system and outputs from the system and defines operating procedures. It has been designed to be easy to learn and is adaptable to the level of skill of persons operating and maintaining electronic switching systems.

3.8.3. Compilers

When installed in a computer (*host* machine), a compiler converts a program written in a high-level language to a program written in code compatible with the machine on which the program is to run. The high-level language program is written by a programmer and is known as the *source,* source code, or source program. The machine on which the program is to run is known as the *target* machine and the machine code program is known as the *object,* object code, or object program. Source code contains all of the comments inserted by the programmer to help comprehension of the construction of the program. It is readable and understandable by analysts and programmers. Object code is the string of bits which must be fed into the target machine so that it will perform as required. It contains bits to initialize registers, allocate parameters and storage, activate operating system and supporting features, etc. It is only understandable to persons intimately familiar with the target machine.

In making the transformation from source code to object code, the compiler performs several sequential tasks. A lexical analyzer identifies characters and character strings, and groups them into *tokens*. A syntax analyzer groups the tokens into structures denoting statements, identifies errors, and converts token groups to an intermediate code. A semantic analyzer verifies the integrity of the process. The compiler optimizes the intermediate code producing a representation from which an efficient target program can be derived. Finally, a code generator produces object code which will run in the target machine to achieve the objectives set for the source program.

3.8.4. Switch Software

The software required to develop, operate, and support a modern digital switching system may consist of one to several million lines of code. The unique character of the task is reflected in the need to develop a generic package which can accommodate installations of widely different sizes, provide various levels of custom performance, and manipulate constantly changing data bases so as to serve the individual requirements of unique subscriber populations. Each installation will be modified as needs change, usage grows, numbering plans expand, new routes are introduced, functions and features are added, charging parameters are altered, and new technology becomes available.

The *on-line* software which controls each installation is a unique package of code generated from an office-independent program, which provides the capability of performing all of the functions defined for the system product; and office-dependent data, which describes size, numbering, routing, and arrangements of lines and trunks, and information concerning each subscriber. It consists of five major parts: executive, administrative, fault processing, diagnostics, and call-processing. The *executive program* provides basic timing, allocates resources (activates program tasks, allocates memory, establishes priorities, etc.), and controls data transfers between processes and input/output devices. The *administrative program* updates and modifies the resident data bases describing network, office, and customer parameters under the control of maintenance personnel. The *fault processing program* responds to malfunction alarms, determines the source of the fault, decides whether and how to reconfigure the system, and restores normal service. The *diagnostic program* isolates the cause of a fault so that the system may be repaired. It is activated automatically, or on the command of maintenance personnel.

The *call processing program* directly controls switching activities. In response to requests for action initiated by the customer, or other facilities, the program generates commands which reflect the state of the call and the action required to advance it to the next state. Action requests are detected by scanning the switch terminations (customer loops and trunks) and comparing the present state with the state on the previous scan. A change signals that action of some sort is required. The decision on what action is appropriate can be made in several ways. In *time-division* decision making, the processor is applied to situations in sequence and generates a suitable command for each. *Function-division* decision making divides the processor into a set of specialized units which respond to specific situations. They are invoked as needed. *Call-division* decision making employs several identical processors which can handle all steps in a call. When an event occurs, the call processor assigned evaluates the event with reference to the current state of the call, and then takes appropriate action on the basis of the present state of the call and the event.

When the processor receives a call request, it must find a path through the switching network which can be assigned to the call. Generally, this is done by seizing a free channel in the central portion of the switching matrix and finding a path through the time and space switching units to the calling termination. When digits are received which signify the destination of the call, the path is completed from the central channel through the remaining switching stages to the proper output termination. This connection is maintained for the duration of the call. It is stored in memory (map-in-memory) to record that the specific time slots and space channels are in use and unavailable for other calls.

The tools and data necessary to write the generic program, assemble the office data, compile the on-line software, and test the system are known as *support* software. The relationship between the generic program, office-dependent data, on-line software, and support software is shown in Figure 3.23.

All manufacturers agree that the largest expense in the development of switching systems is the development of the software. Despite the use of techniques described above, the initial product is far from trouble free and must be maintained and enhanced throughout its lifetime. Some of these difficulties are due to the practical impossibility of testing an entire package of software at the ultimate line size of the system. The result is that testing and fixing must be

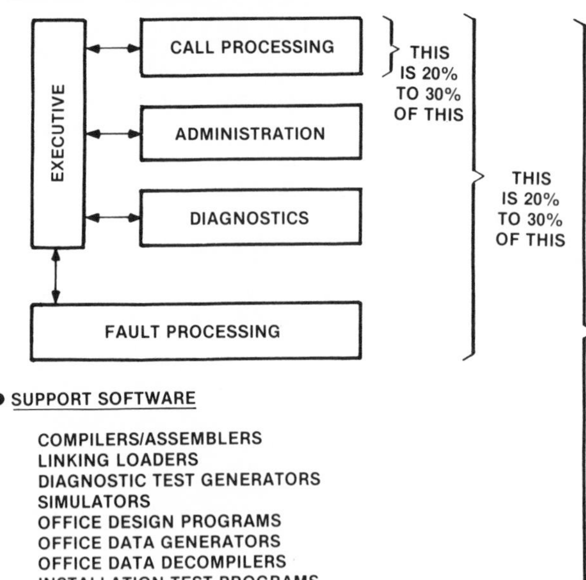

• OFFICE INDEPENDENT PROGRAM (OIP)-GENERIC
• OFFICE DEPENDENT DATA (ODD)
• ON-LINE SOFTWARE (OIP + OOD)
EXECUTIVE
CALL PROCESSING
ADMINISTRATION
DIAGNOSTICS
FAULT PROCESSING
THIS IS 20% TO 30% OF THIS
THIS IS 20% TO 30% OF THIS
• SUPPORT SOFTWARE
COMPILERS/ASSEMBLERS
LINKING LOADERS
DIAGNOSTIC TEST GENERATORS
SIMULATORS
OFFICE DESIGN PROGRAMS
OFFICE DATA GENERATORS
OFFICE DATA DECOMPILERS
INSTALLATION TEST PROGRAMS
ETC.

Figure 3.23. Major components of on-line and support software associated with a digital switch.

done in the field as a succession of increasing line size units are installed. While individual programs and modules can be tested and analyzed thoroughly, it is message handling between programs (and programs and databases), and other activities driven by the offered traffic, which cannot be simulated to the ultimate level. These areas must be carefully monitored through a series of field trial installations to determine if the software has the real-time capability to handle enough simultaneous events.

3.8.5. Next Generation Software

In common with other areas of high technology, software is an active field of research and development in which the concepts, tools, and applications are undergoing change at an increasing rate. The application of new software techniques depends upon the scale and architecture of available processors: in turn, the scale and architecture of available processors is influenced by what can be done in software. For many years, all machines have been designed to employ sequential processing, that is, they operate on one word of data at a time, even when the same operation must be performed on thousands of words. (They are called von Neumann machines to recognize that they are derivatives of the original design by the father of modern computing, John von Neumann.) In this period, software has employed higher and higher level languages in which programmer understandable statements correspond to an increasing number of machine instructions executing sequential actions. Software engineering has been introduced as languages became available to implement and administer the concepts.

In the early years of the 1980s, organizations in Japan, the United States, and Europe inaugurated major research programs in information technology directed towards the *fifth generation* of computer systems which will employ new architectures and new software concepts.(9) They will incorporate parallel processing, and exhibit the ability to make judgments, to draw inferences, and to perform seemingly intelligent functions. While it can be argued that none of these abilities are essential to switching, it is inevitable that success in these areas will have an effect on new systems.

3.9. Artificial Intelligence

The idea of creating machines which can perform functions similar to those performed by human beings has engaged the attention of many persons. Since the early 1960s, to create artificial intelligence (AI) has been a serious goal of modern computer science. Much of this work has been directed to the writing of programs which attempt to manage specific situations through the use of inferences formed by comparison with an existing knowledge base in the fashion illustrated in Figure 3.24. What is to be achieved is defined by the *goal* provided to the *inference generator,* a unit which compares specific information concerning the situation with knowledge of previous situations, deduces relationships, and provides specific directions for use in the present situation. Because it is concerned with *learning* and the acquisition of knowledge, the lower half of Figure 3.24 must be established and exercised before effective direction can be given by the system. During the learning process the inference generator is *trained* to construct correct relationships: during the application process the inference generator receives feedback from the situation being managed so as to modify its directions to satisfy the required goal.

How to *train* a machine is one of the difficult areas of AI research. Ideally, during learning, the machine will organize relevant information into a whole, recognize areas which require more information, seek and acquire gap-filling data, extract laws or relationships governing the area, use them to infer other information (rules of thumb) to solve the problem (such as maximize performance

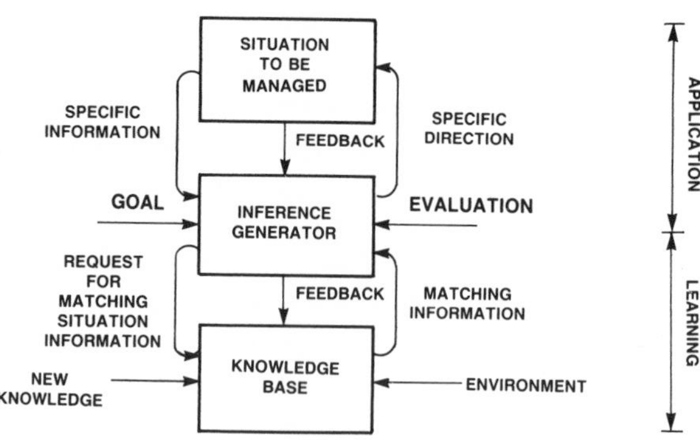

Figure 3.24. Simple model of intelligence applied to management (optimize, prevent, stabilize, etc.) of a specific situation.

of a task), and give directions accordingly. Techniques at the level of training individual logical actions are known, but the application to a complex system is not.

How to represent knowledge so that the interrelation of information is apparent and connections can be made for various purposes is another difficult area of AI research. In machine form, knowledge consists of data structures that may describe objects, previous events, and task breakdowns. Interpretive procedures are used to identify specific objects in the domain of the application, related events that have taken place, and sets of tasks which must be performed to achieve the required goal. Each must be placed in the context of the environment of the specific situation. Linking this information together forms the specific *knowledge base* with which the inference generator must work to extract *rules* that can be applied to the particular situation. Each new piece of information must be assimilated in the knowledge base—it must be interpreted, formatted, and nested with existing information so as to continue the process of linking.

Machine implementation of the model of Figure 3.24 has not been achieved. When it is, it will provide computer-based *intelligent assistants* for most professional activities including the engineering, operation, administration, and maintenance of telecommunication systems. In the meantime, systems of more limited capabilities will begin to assume much of the logical analysis associated with these functions. They are called *expert* systems.

3.9.1. Expert Systems

Expert systems are software programs which solve problems in limited, well-defined areas on the basis of incomplete or uncertain information. They mimic the reasoning power and knowledge of human experts, who are often required to make decisions without all the facts using their *feel* of the situation and rules of thumb which reflect previous successes. Expert systems differ from conventional programs in that the solution is not directly computable because not all of the tasks involved in the solution can be defined precisely. For these tasks exact algorithmic relationships are not available.

In a limited domain, expert systems are constructed to contain the special knowledge, judgment, and experience of a human expert. This activity is illustrated in Figure 3.25. The human expert is interrogated by a person skilled in computer science who translates the expert's descriptions of the tasks involved, their sequence, the laws governing subtasks in which information is generally complete, the rules of thumb which bridge the areas in which information is lacking, the likely areas of iteration, and what constitutes an acceptable result. Known as a *knowledge engineer,* the computer specialist structures suitable modules which contain planning, information, reasoning rules, and knowledge concerning the world associated with the application. The system is tested against the expert and modified until the results are considered satisfactory. Through a natural language interface, the user states the goal of his activity (such as to optimize, stabilize, or prevent certain situations), and provides situation-specific information so as to orient the system (such as physical properties of objects

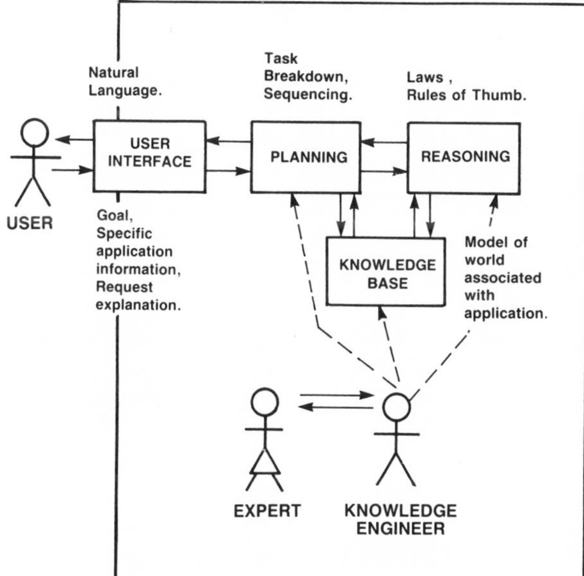

Figure 3.25. Basic structure of expert system.

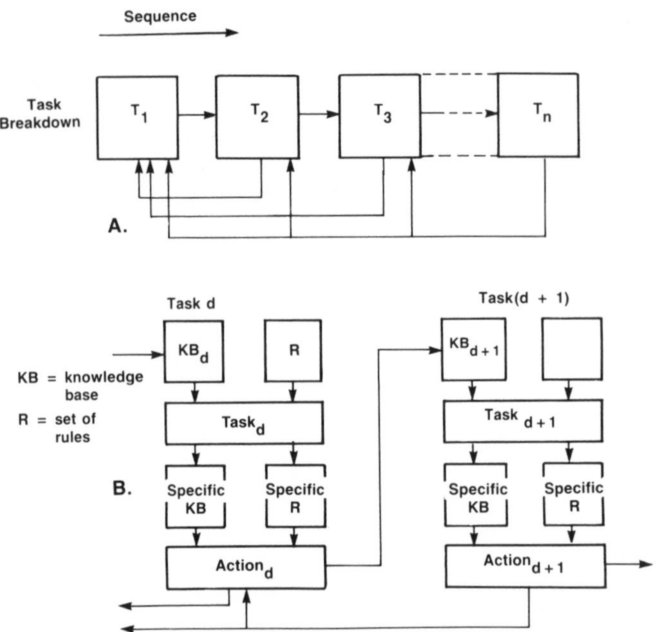

Figure 3.26. Organization of an expert system. The planning module identifies a set of tasks and arranges them in a sequence which will solve the problem, as shown in A. The reasoning and knowledge base modules interact to perform each of these tasks, as shown in B.

involved, any restrictions on solutions, and acceptable levels of risk). The system responds to the user providing directions for achieving the goal and may provide an explanation of the solution if requested. If appropriate, the system may respond with several solutions incorporating different levels of risk or other parameters.

The structure of the expert system shown in Figure 3.25 is amplified in Figure 3.26. In section A, the action of the planning module is depicted as identifying a sequence of tasks, T_1, T_2, T_3, . . . , T_n which must be performed to achieve a solution to the problem posed by the user. At the completion of each task, information is passed forward to the next task, or passed back to a previous task for further iteration to refine the information or to change the information in other ways. Since control of them is relatively simple, a sequence of tasks is preferred. However, some applications may require parallel tasking, and others may require a tree structure or other format. It is the function of the planning module to determine the appropriate topology and break down the tasks accordingly.

The interaction between the knowledge base and reasoning modules is shown in section B of Figure 3.26. At the start of each task, the knowledge base (KB) is increased by the information developed in the previous task. The set of rules (R) contained in the reasoning module remains the same. Specific knowledge and rules are extracted on the basis of the task to be performed. When brought together in the action subtask, they result in information which is passed forward to the next module, or backward to a previous module. In the task which completes the sequence (T_n), the specific knowledge base represents the solution which achieves the goal and the specific set of rules will be those which define a realistic solution. Assuming the application of these rules tests correctly, the action will be to furnish the specific knowledge base to the user as fulfilling the goal for the specific situation stated.

When well-designed, expert systems are highly modular: the nature and sequence of the tasks to be performed, the reasoning rules, and the knowledge base are separated from one another so that additions and refinements can be made in each area without affecting the others. Rule-based systems can be as good as the combination of human expert and knowledge engineer can verbalize and represent the various expert processes involved. What is more, such systems will provide a good solution every time and can be used by many persons simultaneously. They represent an important opportunity to multiply scarce technical resources.

3.9.2. Speech Recognition

Communicating with a machine requires a person to use those things which are discernible by it, such as keyboard imprints, or else the machine must be fitted with special interface equipment which can recognize what the person says to it. Verbal communication is inextricably entwined with human intelligence. To perfect it requires years of practice during which information on sound, spelling, grammar, structure, and context is absorbed. Understanding the role of these intellectual activities, and interpreting them for use by a machine, is a formidable task. Considering the complexity of speech recognition, it is not surprising that the development of a machine to perform the combined functions of ear and brain has not been achieved—yet!

Recognition is based upon extracting a set of features by processing the electrical signal representing an isolated word in some way, then matching the feature set against a library of *templates* representing the words which have been *taught* the recognizer. Most likely, these templates will have been formed earlier by one speaker, or several speakers, speaking the individual words to the recognizer. During the training period the machine may use several examples of the same word (spoken by one person, or several persons) to compute an average template. The features used may consist of spectral information (frequency vs. amplitude) drawn from various frequency bands occupied by the original sounds, or may be the set of linear prediction coefficients (LPCs) which best fit the word. Because speaking rates vary, matching the features of the spoken word against the existing templates may require further processing in order to achieve time

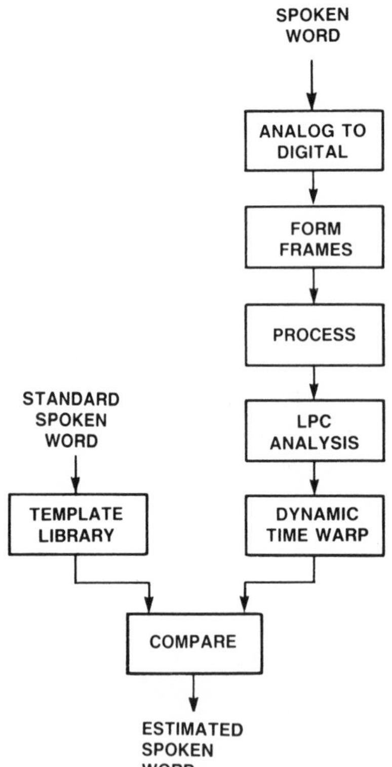

Figure 3.27. Block diagram of LPC-based word recognizer. On the basis of comparison with a set of standards, the system estimates which word was spoken.

alignment with the reference patterns. This process is known as *time warping*.[10] The principle of an LPC-based word recognizer is shown in Figure 3.27.

Isolated word recognizers can identify 10 isolated words spoken by a few thousand persons, and 1000 isolated words spoken by one person, with an average error rate of around one in ten. In a recent test, over 2000 persons spoke more than 11,000 digits over telephone circuits. Using an LPC-based isolated word recognizer trained with recordings of the telephone voices, an average recognition of 93% was achieved.[11] The principles of isolated word recognition carry over into the recognition of continuous speech. Here, the structure of the statements and the order of the words, as well as the words themselves, may assist recognition. Workers in this area usually employ limited vocabularies and define the phrase and sentence structures which may be used. Connected word recognizers may have a vocabulary of around 100 words which are fitted into restricted formats. Through the use of parsing and semantic analysis the error rate may be as low as one word in 50.[12]

3.9.3. Speech Synthesis

The inverse of speech recognition is speech synthesis—the action of producing intelligible sounds from text input or other cues identifying the words to be spoken. Given that there are some 20,000 English syllables to be reproduced, that each word must be broken down into its syllabic components precisely, and that the sound of the same written word may change with context, the processing task is formidable. An unlimited text-to-speech system is several years away. However, units which are restricted to a limited vocabulary and syntax are available today as voice answerback and announcement systems, or in games.(13)

3.9.4. Automatic Programming

Employing a well-defined language and constructs, programming is an intellectual pursuit that results in sets of coded instructions for the direction of machines. Automatic programming depends on the exact specification of what has to be accomplished, the ability to express it in a language which can be processed by a computer, the exact application of a myriad of data for the correct purpose at the appropriate time, and artificial intelligence equal to stepping through whatever constructions are necessary to complete the task. The marriage with data is very difficult; nevertheless, work on semiautomatic *programming assistants* with embedded knowledge is making some progress. In the meantime more evolutionary approaches are receiving attention. New languages which allow the user to state what must be done, not how to do it, are becoming available. Called nonprocedural languages, they emphasize *data* flow as opposed to the *control* flow of conventional languages. In a data flow language, the relationships among the data groups are used to determine the order in which procedures are executed. By analyzing the program statements, the compiler determines the sequence of operations to be performed and causes them to be executed when data are available. These efforts are directed at producing capabilities which can generate the entire program given a simple statement of the objective to be achieved. They hold out the promise of eliminating much of the labor of applications programming and of improving the reliability of software products.

3.10. Signaling and Protocols

Intercommunication requires the cooperation of sending and receiving units, and facilities in between. Information must be passed from the sender to the intended receiver in order to establish a connection over which messages can flow. In telephone networks, this action is known as *signaling* and the information is known as signaling information. It directs the progress of the call, provides address information from which the required connections can be made, and reports on the status of the connection process. In data networks, somewhat similar functions are performed. The necessary information is often added to the data packets. It is defined by protocols which are associated with different levels of the message transfer process.

3.10.1. Telephone Signaling

Establishing a connection means determining the status of individual sections of the network which can be used to provide a talking path, instructing the equipment to connect idle sections, monitoring the use of the connection, and disconnecting the connection when it is no longer needed. When a caller lifts his telephone—a condition known as going *off-hook*—current flows from the local switch to the telephone. This fact alerts the switch to a request for service. When the caller replaces his telephone—goes *on-hook*, that is—current ceases to flow, and the switch registers a disconnect. This signal triggers a series of events that return the sections of the network to idle status ready to participate in other calls.

Upon receiving a request for service, the local switch connects equipment to the line which applies dial tone, indicating that the caller may send address information by dialing or keying the called number. The type of telephone instrument which will be used is an essential piece of information for the local switch so that it will connect the proper common equipment to accept the codes which make up the called number. With a rotary dial, the digit information is passed by interrupting the off-hook current as the dial returns to its rest position. The number of interruptions (dial pulses) is equal to the digit dialed: the duration of the pulses is short enough so that the supervisory circuits do not mistake them for an on-hook condition, and proceed to disconnect.

With a push-button telephone, the digit information is encoded as a combination of two tones selected from seven tones in the range 697 to 1477 Hz. Multifrequency (MF) signaling is used extensively within the network to control switches and trunk connections. Since the range of frequencies falls within the voiceband, it is referred to as *in-band* signaling. Users of push-button phones can hear the digit tones while keying. For some long-distance connections and most international connections, setting up the call involves other bursts of audible tones. Information on the tones associated wtih supervision, addressing, and reporting in the United States is summarized in Figure 3.28.

Signaling information can also be carried *out-of-band*. That is to say the signaling tones do not fall in the 300–3400-Hz voiceband. Single frequency signaling using 3700 Hz in the United States, and 3825 Hz in Europe and elsewhere, is sometimes employed. It is important to realize that while these frequencies are outside the voiceband, they fall within the 4 kHz channel assigned to each voice circuit and are carried along in the same channel. Another technique assigns a separate channel to serve the needs of a group of circuits and facilities.

		FREQUENCIES
APPLICATION	(Hz)	CADENCE
Dial	350 + 440	Continuous
Station Busy	480 + 620	0.5s ON: 0.5s OFF
Network Busy	480 + 620	0.2s ON: 0.3s OFF
Ring Return	440 + 480	2.0s ON: 4.0s OFF
Off-hook Alert	Howl	1.0s ON: 1.0s OFF
Recording Warning	1400	0.5s ON: 15s OFF
Call Waiting	440	0.3s ON: 9.7s OFF

Push Button Digits	
0	941 + 1336
1	697 + 1209
2	697 + 1336
3	697 + 1477
4	770 + 1209
5	770 + 1336
6	770 + 1477
7	852 + 1209
8	852 + 1336
9	852 + 1477

Figure 3.28. Tones and codes used in North America for signaling and supervision.

This is known as *common channel signaling* (CCS) and is discussed in Section 4.1.2.1. It is implemented in those networks served by computer-controlled (SPC) switches. In North America a growing fraction of the present facilities are served by CCIS (common channel interoffice signaling).

3.10.2. Data Protocols

A protocol is a set of procedures that allows computer-based terminals and machines to establish, maintain, and complete communications through the network that connects them. Protocols are rules for controlling the flow of data between computer-based systems to ensure that:

- the data are delivered to the correct location;
- they contain no more than a specified level of errors;
- they flow relatively smoothly; and
- the actions which facilitate communication are completed adequately.

In voice communication, the participants are able to ensure the completion of these actions without difficulty—probably without giving them a conscious thought. When machines communicate, all of these details (and more) must be conveyed, correct actions taken, and confirmed to the initiating location, before message transmission commences. Communication is required between various levels of the machines at either end to ensure useful activity. To guide the development of computer and data processing networks, the International Organization for Standardization (ISO) has constructed a systems interconnection reference model known as the Open Systems Interconnection (OSI) model. Through its use, services, protocols, and telecommunication system structures can be defined which permit the exchange of data between distributed information systems consisting of equipment supplied by many vendors. Since 1978, CCITT has been engaged in a parallel development. A conscious effort has been made to establish functional equivalence between the ISO and CCITT models, and it is likely that they will be indistinguishable in all important aspects.

3.10.2.1. Open Systems Interconnection (OSI) Model

The OSI Model is shown in Figure 3.29. The seven levels, or layers, define a hypothetical architecture known as *Open Systems Architecture* (OSA), which can be used to formalize the information which must flow from one machine to the other. Of the seven layers, the first four are concerned with the initiation and control of communication. Layers 5, 6, and 7 facilitate the use of resources for the users' applications. Layering divides the communication process into sequential steps of increasing abstraction which can be substantially independent of one another provided the upper layer knows only of the services performed by the lower layer, and does not depend upon how they are implemented. A layered architecture of this sort can accommodate different implementations—as in equipment from different manufacturers—and can accept new implementations—as occasioned by the development of new technology. To ensure success, it is important that each level perform a well-defined, differentiable class of functions, and that similar functions be assigned to the same level.

OSA emphasizes peer level-to-peer level protocols and separates the many functions which must be performed to achieve satisfactory communication between data processors. Layer 1 (Physical) is concerned with the physical connection to the communications medium. It includes the operations of establishing, maintaining, and terminating a connection to the transmission medium. Layer 2 (Data Link) is concerned with grouping input data into frames and transmitting them over the physical link with the necessary synchronization, error control, and flow control. It *hides* the details of different link and physical protocols from higher layers. Layer 3 (Network) is concerned with the structure and format of the message and provides whatever switching and routing information is required to support the connection.

In the real world data communication can occur between two machines over public packet data networks, local area networks, private terrestrial long-haul networks, and packet satellite networks. Layer 4 (Transport) provides end-to-

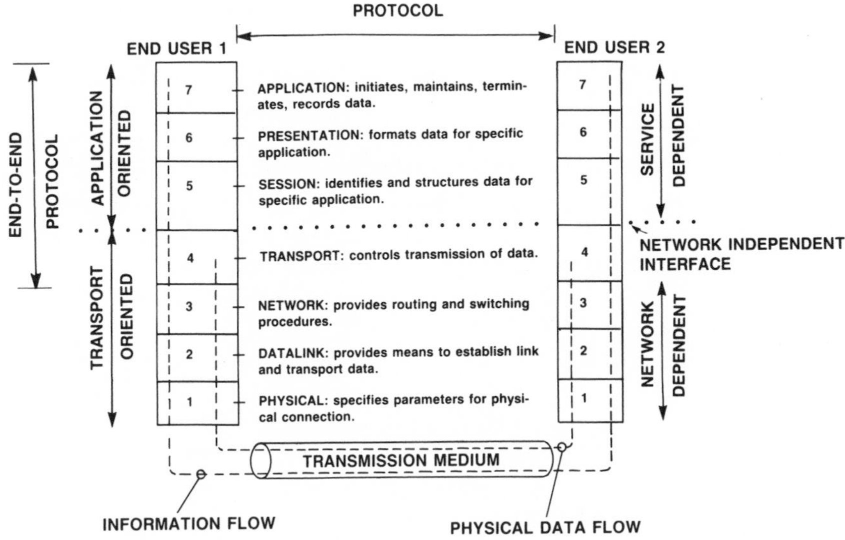

Figure 3.29. Communication protocol hierarchy in ISO Interconnection reference model. The architecture may be referred to an an Open Systems Architecture (OSA) and the model is known as the Open Systems Interconnection (OSI) model.

end data transport service. It originates information on data integrity, duplication or loss, and in-sequence delivery, and provides end-to-end flow control. When necessary, it includes the capability to fragment and reassemble messages to accommodate different packet lengths, formats, and similar requirements which may arise from the use of several different networks to complete the connection. Layer 5 (Session) coordinates the dialog between users, establishing logical connections between them, reliably transferring data, and terminating the exchange upon request. Layer 6 (Presentation) interprets the meaning of the information transferred, managing the formats and transforming the data in accordance with the requirements of Layer 7 (Application). Layer 7 provides system management functions and initiates, maintains, and terminates activities germane to the application specified by the users.

3.10.2.1.1. Level 2 (Data Link) Protocols. In packet data communications systems, the sending node (host computer or terminal) transmits a series of sequentially numbered packets and waits for acknowledgment of receipt. The adjacent nodes receive the packets, test them for errors, acknowledge receipt, and wait for more packets. If acknowledgment is not received within a certain time, the sender *resends* the packets. If communication is not established after a number of resends, the sending terminal notifies the user of trouble. This behavior is implemented in Layer 2 where the protocols are concerned with the flow of information at the lowest level of the network. Their task is to ensure

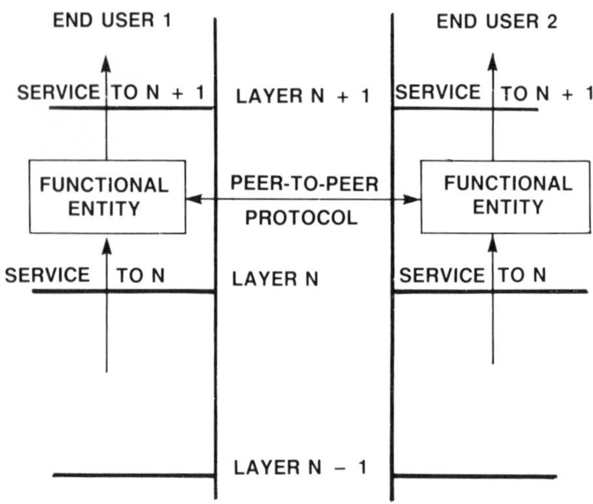

Figure 3.30. Relationship between services and functions in the OSI model.

that all of the information submitted for transmission at one node is eventually received at an adjacent node with nothing missing, and nothing duplicated.

3.10.2.1.2. Layer Concepts. In the OSI model, each layer provides service to the contiguous higher layer by adding *functions* to the service provided to it by the contiguous lower layer. The concept is illustrated in Figure 3.30. Thus, Layer N may perform functions such as error checking—and provide error-free data to layer $N + 1$, or flow control—and provide a continuous stream of data to layer $N + 1$. These functions are achieved by protocols (peer-to-peer protocols) employed between similar functional entities in the users' nodes. They are supported by services from the lower layers. The two functions cited are performed in Level 2. They are supported by the physical connection—the service provided by Level 1, and communicate through the services provided by Level 1. In fact, in general, peer entities in level N make use of the services of level $N - 1$, to pass information. In all but Level 1 it is a *logical* connection.

The seven layers of the OSI model are grouped in different combinations in Figure 3.29. These relations may become more obvious when the domain of use of the layers is identified, as in Figure 3.31 (which is based on Figure 1.1). Layers 1, 2, and 3 are associated with the flow of data from node to node, and are *network* dependent. Layers 5, 6, and 7 are associated with the use to be made of the information exchanged. They are *service* and *application* dependent. The combination of Layers 1, 2, 3, and 4 includes all that is required to pass data between the users. They are *transport* oriented. Layers 4, 5, 6, and 7 are all concerned with *end-to-end* operation.

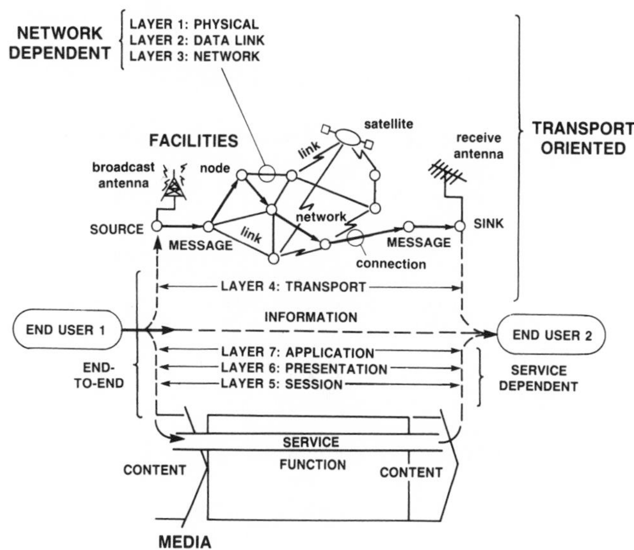

Figure 3.31. The domains of use of the seven layers of the OSI model.

3.10.2.2. *Systems Network Architecture*

The work of ISO and CCITT has yet to be completed. Meanwhile, the concepts contained in the OSI model are influencing the evolution of existing data networks. Over 20 years ago, IBM recognized the need for a master interconnection strategy so that diverse products and applications could share computers and telecommunication facilities. The result is Systems Network Architecture (SNA), an arrangement of networking functions which provides for all levels of interaction from physical and electrical interconnection of devices, to the logical connection of applications. Like OSA, SNA groups similar functions into multiple layers, as shown in Figure 3.32. All of the layers may reside in the user's computer, or the lower levels may be contained in a separate communications controller and the intermediate levels may be contained in a front-end communications processor. The allocation of logical functions is flexible depending on the equipment available, and some functional options can be dynamically assigned at the time a session is established. More information on SNA networks is found is Section 4.7.5.

3.10.2.3. *Other Models*

Two other network architectures will be mentioned briefly. One DNA (Digital Network Architecture) was developed by Digital Equipment Corporation

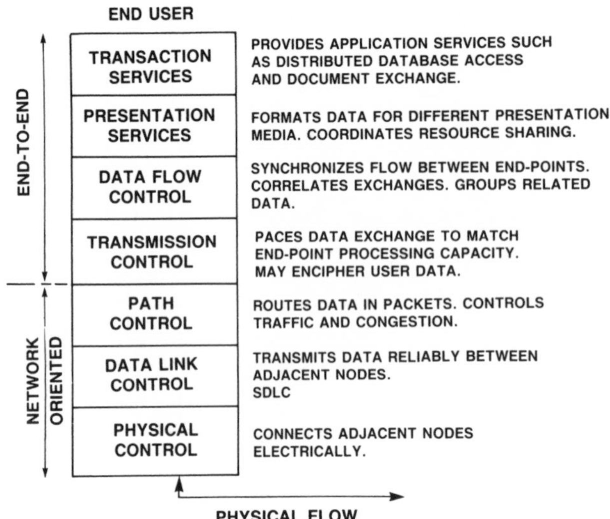

Figure 3.32. Layers in the Systems Network Architecture (SNA) model.

(DEC) to interconnect DEC products into networks (DECNETs). It employs the five-layer structure shown in Figure 3.33. The physical connection is full-duplex and messages are only transmitted in response to a request from the receiver. The other architecture has evolved with the development of ARPANET, the world's first packet data communications network, sponsored by the Advanced Research Projects Agency (now DARPA, Defense Advanced Research Projects Agency), a unit of the Department of Defense. Protocol layers are associated with major hardware and software entities. An interface message processor (IMP) connects a host computer to a packet data communication network, and the IMP-to-IMP protocol establishes a virtual circuit communications path between hosts. It employs the four-layer structure shown in Figure 3.33. Information on the extension of ARPANET to multiple network operation is given in Section 4.7.4.3.3.

C. MICROSTRUCTURES

In this section, we describe the principles of operation and the techniques for making the electronic and optical devices required by advanced telecommunication facilities.

3.11. Solid-State Electronic Devices

Late in 1947, the course of electronics development was irreversibly changed by the invention of an amplifying device constructed from extremely pure, single-crystal, semiconducting material. Known as a *transistor,* early devices were largely based on germanium (Ge). Later devices used silicon (Si) to take advantage of its superior properties. Today, the solid-state revolution is firmly based on silicon technology. Devices constructed on gallium arsenide (GaAs), indium phosphide (InP), etc., fill specialized roles in the spectrum of applications.

3.11.1. Bipolar and Field-Effect Transistors

Transistor action in semiconducting crystals depends on the conduction of electricity by two charge transport mechanisms. One is associated with negatively charged conduction electrons which move freely through the crystal lattice. The other is associated with a deficiency of one electron in some of the valence bonds

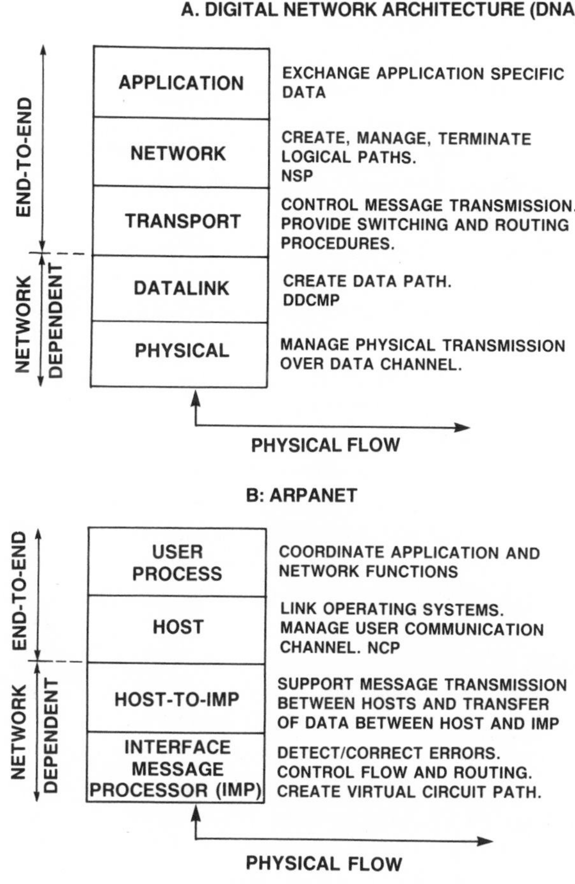

Figure 3.33. Protocol layers in DNA and ARPANET.

forming the crystal. Called a *hole*, it moves from bond to bond as valence electrons move to fill existing deficiencies, leaving new deficiencies behind them. By incorporating precisely controlled levels of impurities in the crystal, either type of charge can be made the *majority* carrier (the other is the *minority* carrier).

In its pure form, silicon contains few free current carriers and is highly insulating. The addition of selected impurities from group III or group V of the Periodic Table produces an excess of free electrons or holes which result in a large increase in conductivity. Elements such as phosphorus (P), arsenic (As), and antimony (Sb)—group V elements—possess five valence electrons, one more than silicon (a group IV element). When one of these elements is substituted for silicon in the crystal lattice, four of these electrons are used to make chemical bonds to the silicon atoms. The fifth electron remains weakly bound to the impurity from which it is easily freed by thermal energy. At room temperature, each group V impurity atom donates one free electron to the crystal, leaving a positive impurity ion bound to its silicon neighbors. The free electrons can move about the crystal and support conduction. Because current is carried by *n*egative charges, materials of this type are known as *n*-type materials.

Elements such as boron (B), aluminum (Al), indium (In), and gallium (Ga)—group III elements—possess three valence electrons, one less than silicon. When one of these elements is substituted for silicon in the crystal lattice, the three valence electrons form bonds to three of the neighboring silicon atoms. A bond to the fourth neighbor is formed by the impurity atom accepting an electron from a silicon atom to form a negative impurity ion bound to its silicon neighbors. The hole produced by the transfer of an electron from a silicon atom to the impurity atom is free to wander from silicon atom to silicon atom in the crystal and support conduction. Because current is carried by *p*ositive charges, materials of this type are known as *p*-type materials.

The operation of silicon devices depends upon the formation of junctions between *n*- and *p*-type material (called *pn* junctions). Two *pn* junctions in close proximity form a *bipolar* transistor (which can be *npn* or *pnp*). The three consecutive regions are known as the *emitter*, *base*, and *collector*. When properly connected, minority carriers flow from emitter to collector under the control of a voltage applied between emitter and base. The collector current flows through an external load where the power developed can be 10,000 times (or more) that absorbed in the emitter–base connection (a gain of 40 dB). Bipolar transistors exhibit high speed, consume more power than other types, and can be used for both analog and digital applications.

In a *field-effect transistor* (FET), majority carriers flow from *source* to *drain* under the control of a *gate*. In one embodiment, a metal gate electrode is separated from the silicon semiconductor material by an insulator such as silicon dioxide (SiO_2). This style of construction is called MOS (metal–oxide–semiconductor) and the transistor is known as a MOSFET. It is the most common member of a class of devices known as IGFETs (insulated-gate FETs). MOS transistors

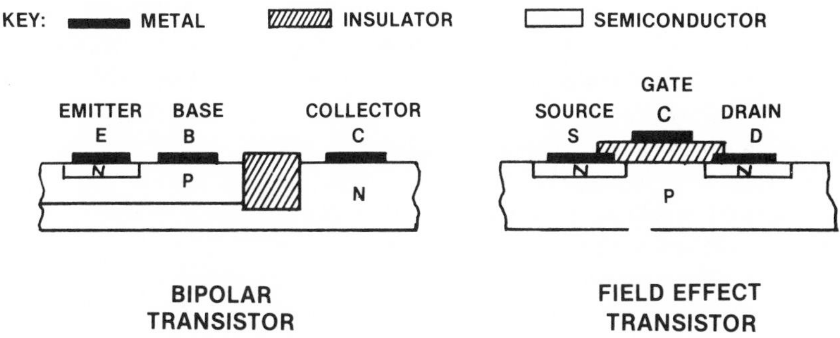

Figure 3.34. Basic construction for bipolar and field-effect transistors.

exhibit slower speed than bipolar transistors, require less power, and are used mostly for digital applications. (They can perform analog functions, also.)

The basic constructions of these elementary transistors are shown in Figure 3.34. On the structures shown, all connections are made from the top surface. While this is not the usual construction for single devices (called discrete devices), it is the configuration which is employed in integrated circuits—by far the most common application.

As the technology has been developed, variations of these basic forms have been perfected to optimize specific performance. For example, bipolar transistors exist as transistor–transistor logic (TTL), Schottky TTL (STTL), emitter-coupled logic (ECL), emitter function logic (EFL), and integrated injection logic (I^2L). They exhibit fast performance with various degrees of propagation delay. MOS transistors exist in *p*-MOS or *n*-MOS versions, depending on the polarity of the majority carrier, and as complementary MOS (CMOS) in which the complementary properties of *p*-MOS and *n*-MOS devices are balanced to yield low power operation. Despite the fact that CMOS is more expensive than *n*-MOS, it is becoming the process of choice for most digital telecommunication, computer, and consumer electronics applications.

Silicon technology has been applied to devices performing analog functions in the low- and intermediate-frequency areas of the electromagnetic spectrum reaching into the lower microwave region. However, with a few exceptions, the demand for these devices has been relatively small (compared to digital circuits) and the development of designs and processing methods has not kept pace with the work in the digital area.

3.11.2. Silicon Integrated Circuits

Had the invention of the transistor resulted in discrete devices only, it would still be an important discovery, but it quickly became apparent that many transistors can be formed on the same substrate, that passive components such as resistors and capacitors can be constructed, and that interconnections can be made. The result is entire circuits integrated together on small squares of silicon which are the very foundation of modern computing and modern telecommunication and which have revolutionized the entire world of electronics. Integrated circuits (ICs) offer low power consumption, high operating speeds, low volume and weight, high reliability, and low cost. With such desirable features is it any wonder that ICs permeate the electronics industry?

Large-scale integrated (LSI) silicon circuits in common use today operate at rates up to 10 Mb/s. Other silicon ICs operate at frequencies up to 100 MHz. Premium silicon ICs may operate at rates up to 500 Mb/s or frequencies up to 2 GHz. Low volume and weight follow naturally from the fact that a completed circuit chip may average 0.5 cm (somewhat less than one-quarter of an inch) on a side. They consume relatively little power because of their low operating voltages (typically ± 5 V) and small current requirements (typically milliamps). As a consequence, the need for heavy power supplies and substantial supporting structures (such as printed circuit cards, chassis, frames, and racks) is reduced significantly. In addition, the intrinsic properties of the solid-state environment, the relatively automatic manufacture of the chip, and the small number of external connections which must be made in later stages of the manufacturing process result in higher reliability. Modern methods employ approximately 5 in. (127 mm) diameter silicon wafers on which several hundred ICs can be made at a time. The fact that wafer diameters can be expected to increase at least another inch or two promises lower costs in the future.

The major design and processing steps associated with the production of an integrated circuit are shown in Figure 3.35. Starting with the required performance, the circuit is designed using computer programs to manipulate standard arrangements of transistors and other components. Performance is simulated on a computer and checked against requirements. When the designer is satisfied that the circuit performs as desired, a layout is prepared by the computer from cells which correspond to the transistors, components, and devices used in the design. It consists of several layers of geometric patterns which correspond to the diffusion, implant, evaporation, and oxidation steps required to manufacture the circuit in planar form on silicon. At the same time, the computer generates testing information from the simulation on which to base tests of the finished product.

The physical preparation of an integrated circuit begins with an ingot of single-crystal silicon. Thin wafers are sliced from the ingot and polished. These form the substrate on which all succeeding operations are performed. Wafer processing begins with the deposition of a thin, epitaxial layer of silicon. This is the layer in which a variety of diffusions and implants produce silicon having the appropriate charge carrier types and densities so as to form transistors and components, and on which insulators and metals are deposited to form connec-

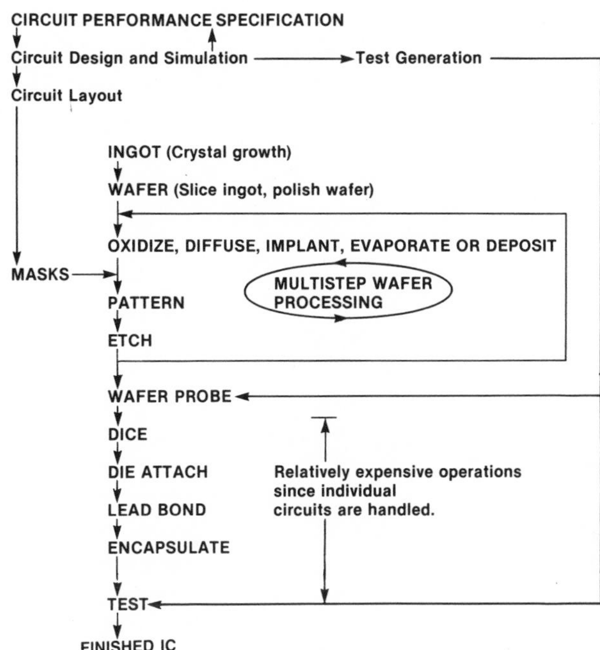

Figure 3.35. Representation of the major design and processing steps required to produce an integrated circuit.

tions. Each operation is guided by photoresist which has been patterned by exposure to light through one of the set of masks designed by the computer. When all masks have been used, processing is complete and the wafer may hold several hundred copies of the circuit. They are tested automatically (usually called wafer probing) and rejects are identified. Since each circuit is processed individually from this point on, a determined attempt is made to eliminate all bad product at this stage. The wafer is sliced (diced) to yield individual dice (or chips). Each good die is attached to a header, leads are bonded from the contact points on the circuit to the header pins, and the mounted unit is encapsulated and tested.

Figure 3.36 shows the increasing complexity and performance of integrated circuits since 1970. Production gate lengths have decreased from approximately 10 μm in 1970 to around 1 μm today. At the same time, the minimum gate delay (a measure proportional to circuit speed) of production circuits has decreased from something over 10 ns to around 0.1 ns, and the capability placed on a single chip has increased almost a thousandfold. In fact, the number of transistors on a chip presently appears to be increasing at a rate between 50% and 100% a year. From 4k RAMs (random access memory chip containing 4000

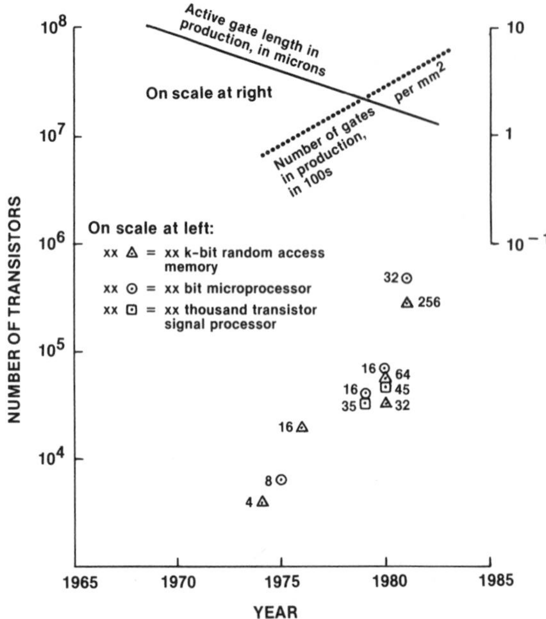

Figure 3.36. Change in integrated circuit parameters, performance and product since 1970. The number of transistors is growing yearly.

bits) and 8b μPs (8-bit microprocessors) in the mid-1970s, it was a gigantic step to 64k RAMs and 16b μPs in 1980, and to 256k RAMs and 32b μPs today. If the last decade was the time of integrated circuits, moving to large-scale integrated circuits, the next decade is the time of very large scale integrated circuits and systems-on-a-chip. These units will require submicron construction techniques and have led to a new area of technology which has been called *microstructure* science and engineering. The minute scale of this technology is significantly affecting the entire electronics industry.

3.11.2.1. Very Large Scale Integration

The relative simplicity of insulated gate FET device performance, geometry, and fabrication technology has encouraged IC designers to seek increasing levels of integration so as to achieve ever greater capability on a chip. While increasing the chip area and improving the efficiency with which FET devices are used has contributed to their progress, the largest contribution has been the decreasing size of the individual structures of which the finished circuit consists.

The progress made in reducing the minimum feature size employed by state-of-the-art production facilities is shown in Figure 3.37. They range from 3 μm in 1978 and 2 μm in 1981, to 1 μm production processes in 1984 and $^1/_2$ μm processes in 1988. Although such dimensions do not approach the ultimate single-line width which may be achieved (perhaps 3×10^{-9} m, or less), many believe that the use of structures with dimensions much less than one quarter of a micron will not be possible because of the buildup of tolerances in the multilevel processing and the vulnerability of such minute features to defects caused by processing, as well as imperfections in the underlying crystal structure itself.

3.11.2.2. Limits to the Scale of Integration

The performance of silicon integrated circuits has increased steadily from year to year. However, it is likely that some physical limits will be reached before too long. Most easily understood is the one already discussed—because of the buildup of tolerances, the minimum dimension of a multilayer structure may never fall below a significant fraction of a micron. Nevertheless, this is sufficient to produce one hundred million, or so, transistors on a chip, and will certainly result in individual memory chips of several million bits and random logic chips of several hundred thousand gates—a number which is far in excess of the total number of gates employed in the logic portions of today's large

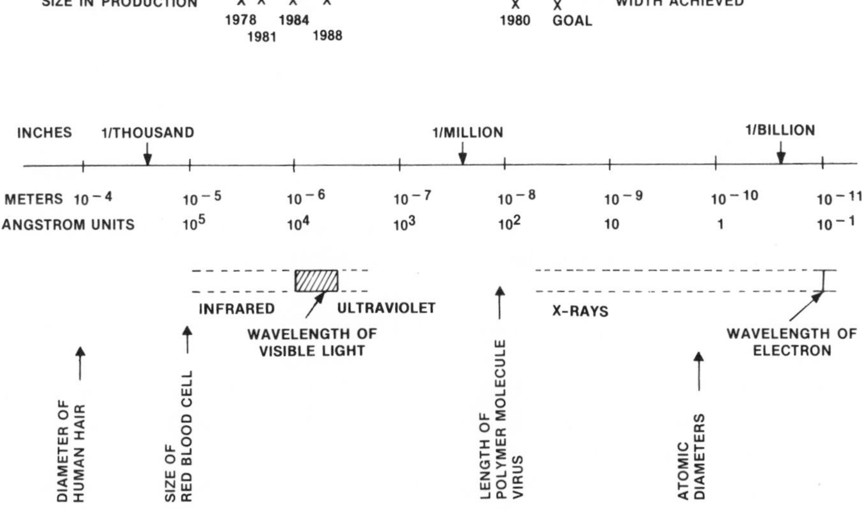

Figure 3.37. Progressive reduction in minimum feature size of production circuits contrasted with dimensions of some familiar things.

computer mainframes. The former depends on a combination of cell design and operation which can clearly differentiate 1's and 0's despite the small active element volumes involved, and the latter depends on achieving functional placement of cells which minimizes the length of the average connection so that delay does not affect speed unduly. The submicron cells must be operated at lower voltage (perhaps 1 V as compared to the present 5 V) to prevent breakdown among the elements. Another factor which will limit performance is the increased propagation delay encountered as the linewidth is reduced. For 1-μ-wide lines this amounts to approximately 1 nanosecond (10^{-9} second) per centimeter and may limit operation to a gigabit, or less. New circuit architectures can be counted on to lessen the impact of this effect. Until recently, heat dissipation was considered to be a limitation. However, microcapillary interface techniques have demonstrated the possibility of conducting away more than enough heat for future systems-on-a-chip.

3.11.2.3. Other Options

Creating planar, submicron structures on individual chips with spectacular speed and processing power is not the limit of possibilities for integrated circuits. At least two operations exist which can further increase the flexibility and scale of integration, albeit at the expense of more complicated processing. They are *wafer scale integration* and *three-dimensional integration*.

3.11.2.3.1. Wafer-Scale Integration (WSI). In wafer-scale integration (WSI), not all chips on a wafer are the same and they are connected to other chips on the wafer to create a total system. Some circuits are duplicated or triplicated for reliability, other circuits test the operating modules, and other circuits rearrange interconnections to optimize functions as required. Interconnections between circuits can be significantly wider than the micron and submicron lines used within circuits, thereby reducing intercircuit delay and transmission loss.

3.11.2.3.2. Three-Dimensional Integration (3DI). Instead of the individual circuits being spread out over a wafer, they can be fabricated on top of each other to form a three-dimensional structure. The possibility of achieving up to ten levels has been suggested. This construction has the advantage that connections between circuits can be shorter than in WSI, but the individual circuits (particularly the first) must be very robust to withstand the stress of processing over and over again as the circuit layers are built up. A natural extension of 3DI (three-dimensional integration) leads to the concept of three-dimensional WSI. Perhaps there will be a need for such complexity—right now, however, developing techniques to satisfy the processing requirements of VLSI, WSI, and 3DI would seem to be more than enough to occupy the entire resources of the world's integrated circuit builders.

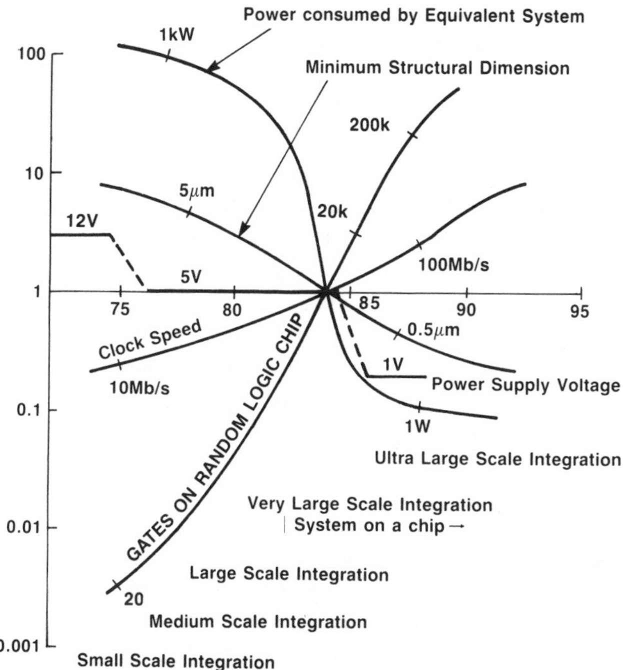

Figure 3.38. The performance of production silicon integrated circuits has improved year by year. Less power and greater speed are a direct consequence of the reduction in feature size. The number of gates on a random logic chip is paced by the ability to design these circuits, not to make them.

3.11.3. Commercial Considerations

3.11.3.1. The Promise of VLSI

In Figure 3.38, some of the parameters affecting silicon integrated circuit performance are depicted as functions of time. Four of the curves show the variation of structural dimensions, number of random logic gates on a chip (devices of this sort are known as gate arrays), clock speed, and power supply voltage. They reflect the progress from small-scale through medium-scale, to large-scale and very large scale integration, and the ultimate density which some call *ultra-large-scale* integration (ULSI). The fifth curve shows the reduction in power consumed by a typical system which can benefit from these changes. As a system of the mid-1970s, it may have consisted of a rack of equipment incorporating small-scale and medium-scale integration, and consumed around 1 kW. As a system of the late 1980s, it will be on a single chip and consume around 1 W. Such is the promise of integrated circuit technology. When fully realized, it will be the keystone supporting the wholesale expansion of computer, commmunication, and consumer electronics applications under the influence of the overwhelming logical capability such systems will support.

3.11.3.2. *Investment*

Achieving this level of miniaturization and performance has required the investment of billions of dollars in machinery, facilities, computer programs, and engineering experience, by leading manufacturers around the world. The demand for computer, communication, and consumer goods had been sufficient to justify, support, and vindicate the commitment of such resources. With 1-μm processes extant and submicron processes a definite future option, new machinery and facilities are vital for the manufacturer who wishes to remain at the leading edge of the market. A fully equipped factory may cost several thousand dollars per square foot, so that to build and equip a modest facility can require the expenditure of $50 million.

But the buildup of expenses does not stop with facilities and machinery—the effort required for definition, design, and layout of a chip has grown exponentially over the years so that the implementation of a moderate system on a chip may require 50 to 100 man years and the software can account for a similar amount of effort. Thus, progress toward VLSI and ULSI and the promise of systems on a chip can only be achieved by industrial organizations with access to major financing sources and an understanding of the requirements likely to result in systems on chips. The recognition of these points is turning systems and equipment manufacturers into semiconductor manufacturers, and semiconductor manufacturers into system and equipment manufacturers.

3.11.3.3. *Government Programs*

So great are the intellectual and financial requirements for full exploitation of V/ULSI, and so important are the potential results for national security and economic survival, that governments have taken a hand. Thus, in Japan, in the late 1970s, the MITI (Ministry of International Trade and Industry) sponsored an industrial program to advance the state of the art of VLSI. In the United States, the Department of Defense (DoD) has long had a goal of achieving molecular electronics. In the early 1960s, government funding resulted in the exploitation of integrated circuits in weapon systems. Starting in the late 1970s, government funding is being applied to the development of very high speed integrated circuits (VHSICs), to the development of micron and submicron processing technology, and to Ultrasmall Electronics Research (USER), a coherent, long-range, multidisciplinary research program directed to the scientific aspects of nanometer electronic structures—to generating a new science of ultrasmall domains. In the VHSIC program, a figure of merit has been defined as the product of the clock rate (the basic speed at which operations occur) and the number of gates per square centimeter. Minimum values of 5×10^{11} gate Hz/cm^2 are expected for circuits employing 1.25 μm features and 1×10^{13} gate Hz/cm^2 for submicron circuits. Applications will extend to weapons systems, satellites, avionics, radar, undersea surveillance, electronic warfare, signal interception, and command, control, and communication systems.

3.11.3.4. Telecommunication Applications

The direct benefits of VLSI/ULSI digital circuit technology to telecommunication will be apparent in terminals and switches—in the nodes and ends of the network. The increased capability and decreased cost they will bring to these points will have an effect on the transmission links. The ability to construct circuits with a million or more transistors on a chip provides the opportunity to place logic and intelligence at almost any point in the network, and to incorporate sophisticated signal and message processing into every user interface. In addition, the sheer logical power which can be contained on the larger chips will surely equal all that is needed by an end office. While an end office on a chip is not very likely for a number of reasons (security, power, connections, etc.), an end office in a small equipment rack is certain. Switching digital multiplexed streams carried by optical fibers over the last few feet to a miniature terminating frame, the chips will handle multiservice traffic in ways which are based on computer science. Costs must decline, service offerings must increase, and the ability to support all of the advanced message requirements of residence and business telecommunication will be assured. Obviously, there is a way to go, yet present technology lends credence to the goal. Contemporary chips containing 32-bit central processing units (CPUs) with random access memories (RAMs), read-only memories (ROMs), and special high-speed logic units (such as multipliers), are already substituting for today's *black boxes*.

3.11.4. Next Generation Systems

The complexity and expense of developing software and hardware for future systems has focused attention on means to create them automatically. The development of the fifth-generation computer, and the possibility of mimicking human intelligence, have stimulated concepts of a *system* compiler which processes the system specification to produce instructions for the manufacture of a chip and code with which to operate the system. An enormous undertaking—yet one which seems intellectually possible and may be achievable if the promises of artificial intelligence can be captured and reduced to practice.

Encouraging progress is being made in the automation of both hardware and software design tasks. Simulation and layout of today's chips is computer aided and hardware description languages (HDLs) are being developed. They will serve as the input to silicon compilers, that is, computer tools which reduce designer understandable circuit descriptions (in HDL) to code for use by chip manufacturing machines. The design of today's software consists of the decomposition of the software specification into ever smaller and more specialized domains until each segment can be executed in a high-level language and compiled into machine code. The development of machine processable specification languages holds the promise of automating this decomposition and of providing means for checking the completeness and integrity of the specification. With perseverance and the incorporation of the reasoning power of artificial intelligence there seems to be no intellectual impediment to the automation of the design of hardware and software to a significant degree—in fact, to a level which combines just the right mixture of automation and personal interaction so as to produce useful, efficient, tested designs.

The development of hardware and software specifications from an overall functional specification describing what the system is to do is probably going to be more difficult. It will depend on the invention of a machine-processable, functional specification language which can serve as input to intelligent systems (expert systems) capable of making architectural decisions, of evaluating the cost and convenience of hardware/software tradeoffs, and of forming inferences from previous systems to guide the present design. When available, it could point the way to efficient evolution of existing products, or make every situation an application for a highly specialized custom design. More than perseverance will be needed to make this happen.

3.11.5. GaAs Integrated Circuits

Materials other than silicon exhibit semiconducting properties and can be used as the basis for integrated circuits. However, none developed so far is as easy to process and manipulate. Nevertheless, for specialized applications, the additional processing complexity yields sufficiently better performance to make the effort worthwhile. Such is the case with gallium-arsenide (GaAs), which is capable of producing integrated circuits whose speed times power products are four or five times greater than those achieved in silicon circuits.[14] Its material properties, operating temperature range, and other parameters, make GaAs suited to high-performance circuit applications. GaAs ICs are produced using planar fabrication techniques which are similar in concept to those employed for silicon ICs. The basic building block for digital applications is a field-effect transitor known as MESFET (metal–semiconductor FET) in which superior performance is achieved using dimensions which are smaller and tolerances which are tighter than those employed in silicon. Starting in the mid-1970s, digital GaAs IC technology has developed at a rate in which the number of gates on a chip has increased by approximately three times every year.[15] Circuits with as many as 10,000 gates have been fabricated.[16]

GaAs also finds application in radios and equipment operating at frequencies above a few gigahertz. While the patterns are not particularly complex and the density is not so great, manufacturing GaAs microwave devices requires as much sophisticated materials technology as GaAs digital circuits. GaAs FET amplifiers generate several watts at microwave frequencies with low distortion and high linearity. They can also be used as high-efficiency microwave oscillators. With an additional gate—to form a dual-gate FET—GaAs FETs can be used as mixers, switches, limiters, and discriminators. Integrating combinations of these devices on a single chip has produced chip sets for microwave radios. Known as MMICs (monolithic microwave integrated circuits), these circuits are built on epitaxial GaAs using ion-implantation techniques to produce FETs and diodes. Special interconnection techniques have been developed to provide low ground impedance and to control parasitic elements—essential features of devices which operate at several gigahertz. Present MMICs include millimeter-wave FET amplifiers, mixers, and broadband IF amplifiers. Ultimately, they will permit the construction of microwave switching matrices, phase shifters for phased-array antennas, and other radio-frequency devices essential to the full exploitation of telecommunication through future advanced satellites. These MMICs will include both analog microwave and high-speed digital circuits on the same chip.(17)

3.11.6. Josephson Junctions

Over 20 years ago, Brian Josephson demonstrated high-speed switching which required virtually no power in devices cooled below their superconducting temperature—that is, the temperature at which the electrical resistance of the

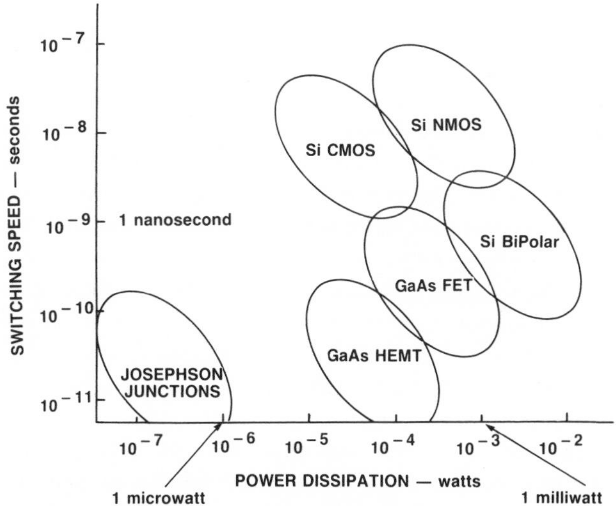

Figure 3.39. Operating regimes for a Josephson junction, GaAs high-electron mobility transistor (HEMT), GaAs field effect transistor (FET), and family of silicon devices.

elements becomes zero. Today, with respect to switching speed and device power, Josephson junctions (JJs) outperform both GaAs and Si devices by one and two orders of magnitude, respectively. A major drawback to the implementation of systems which employ these components has been the availability of equipment to cool them adequately. The development of reliable helium cryostats has removed this difficulty, and JJs are routinely cooled to 4° above absolute zero (−269°C) by immersion in liquid helium. At this temperature they switch in a few picoseconds (10^{-12} s) and consume around one microwatt (10^{-6} W). Josephson technology is unique and not compatible with the physical properties of Si and GaAs devices[18] so that products employing combinations of these technologies are not likely to appear. In addition, material and manufacturing problems are proving less tractable than expected, so that JJs may remain only a laboratory phenomenon.

In Figure 3.39, switching speed and power consumed are plotted for the three families of components. It should be noted that silicon technology has been steadily encroaching on the application area of gallium-arsenide, and that gallium-arsenide is steadily encroaching on the domain of Josephson junctions.

3.12. Solid-State Optical Devices

Within the last few years, with the invention and reduction to practice of low-loss optical fibers, robust optical fiber cables, long-life optical sources and detectors, and splices and connectors, optical telecommunication has become a reality. The propagation of optical energy is governed by Maxwell equations. In this representation, the total energy is divided into modes which are the particular solutions that satisfy the boundary conditions of the situation analyzed. For arrangements in which the device dimensions are very much larger than the wavelength(s) involved, and the boundary conditions permit the propagation of many modes (multimode), the results are synonymous with those derived from the use of classical geometric optics. For arrangements in which distances are comparable to the wavelength(s) involved, and the boundary conditions permit the propagation of only a few, or one, mode (single-mode), the results reflect the wave nature of the energy. As we shall see, multimode and single-mode operation provide different levels of performance and demand different component structures.

3.12.1. Optical Fiber

An optical fiber is a fine strand of optically transparent material (about the diameter of a human hair) in which the refractive index varies from the center to the outside in such a way as to guide optical energy along its length. Transmission of information over an optical fiber has several fundamental advantages over transmission along a copper wire. Optical fibers are electrical insulators: they provide electrical isolation between transmitter and receiver. Optical energy is unaffected by electrical radiation: optical communication can occur in noisy electrical environments without interference. When launched properly, all of the optical energy is guided along the fiber, and interference (cross-talk) from adjacent fibers is nonexistent. Optical frequencies are very high compared to the message bandwidth: optical fibers can be used to transmit wideband signals. The major disadvantage is that optical fibers do not conduct electricity and thus it is impossible to power equipment such as telephones or repeaters down the fiber.

3.12.1.1. Fabrication

Optical fiber intended for telecommunication is generally 125 μm (approximately 0.005 in.) in diameter. At the center is a *core* of 5 to 50 μm diameter which has a slightly higher refractive index than the surrounding *cladding*. This is achieved by producing a radial variation of the material composition during fabrication. Several techniques to do this have been perfected.

The principle of the outside vapor deposition (OVD) process is illustrated in Figure 3.40A. Small particles of very high purity glass formed by reacting a mixture of vapors of appropriate materials are deposited on a rotating ceramic rod. This glass soot is gradually built up into a porous preform along the length of the rod by moving the flame (or the rod) to and fro. As the preform thickens, the chemical composition is varied to produce the desired optical properties. When deposition is completed, the preform is removed from the center rod and sintered to form a glass rod. This solid preform is suspended upright in a ring furnace, the tip is melted, and fiber is drawn. The optical properties of the fiber mimic those of the preform, including radial variation of the refractive index.

The principle of the modified chemical vapor deposition (MOCVD) process is illustrated in Figure 3.40B. A pure quartz tube which will later become the fiber cladding is rotated over a flame which moves to and fro. Suitable vapors flow through the tube causing glass soot to build up on the inside of the heated tube. When deposition is completed, the tube is heated to a higher temperature and collapsed to form a solid preform from which fiber is drawn as described previously.

The principle of the vapor-phase axial deposition (VAD) process is illustrated in Figure 3.40C. Forming and sintering the porous soot preform are completed in one vessel. The solid preform is converted to fiber in the manner described previously. All three of these vapor deposition processes are capable of making preforms from which first-class fiber can be drawn.

A different approach to glass fiber making is illustrated in Figure 3.40D. Twin crucibles are filled with core and cladding glasses, and the combination

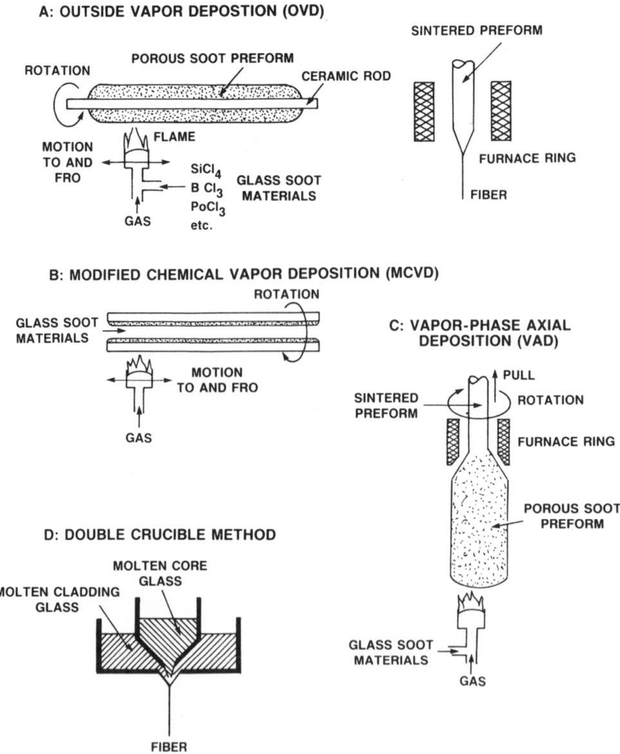

Figure 3.40. Basic techniques for preparation of telecommunication quality optical fibers.

drawn from the bottom where the two flow together. In contrast to the three previous methods in which the length of fiber is limited by the volume of the preform, this technique is capable of making fibers of any length.

Pulling long lengths of high-quality fiber requires first-class, precision machinery with a high degree of computer control and a clean working environment. Fiber diameter is determined by the speed at which the fiber is pulled, or the size of the crucible outlets. Within a few feet of the molten tip of the preform, the fiber is coated with plastic to protect the virgin glass surface from damage before it is reeled on a spool.

3.12.1.2. Physical Properties

Refractive index profiles for four types of fiber are shown in Figure 3.41. They represent the general types in use today. Example (a) is a step index fiber in which the refractive index of the core is typically 0.5% greater than that of the cladding. Example (b) is a graded index fiber in which the refractive index

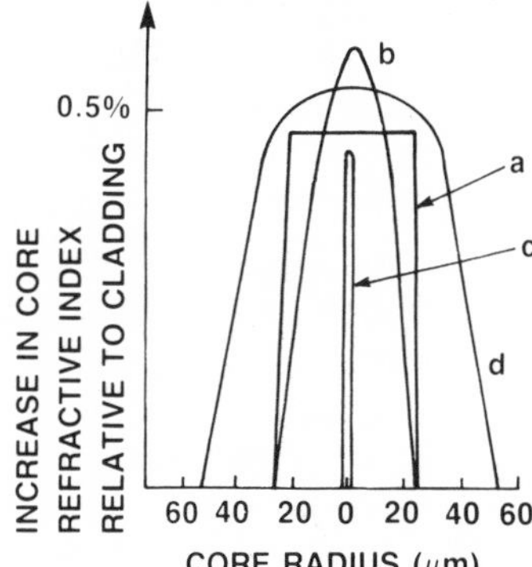

Figure 3.41. Representative refractive index profiles for four categories of fiber. The outside diameter of examples a, b, and c is 125 μm. The outside diameter of example d is 250 μm.

of the core increases in an approximately parabolic fashion to a maximum of some 0.5% greater than that of the cladding. Example (c) is a step index fiber with a very narrow core, and example (d) is a large diameter graded index fiber.

The basic mechanism of transmission of optical energy along cores whose diameter is many times the wavelength can be illustrated using classical optical concepts and is shown in Figure 3.42. For the simple case of the step index fiber illustrated, rays which impact the core–cladding interface at less than the critical angle of incidence emerge into the cladding and are lost from the transmission path. Rays which impact the core–cladding interface at more than the critical angle of incidence are trapped within the fiber and propagate along it. Rays which enter nearly parallel to the fiber axis travel a shorter distance than the rays which bounce back and forth across the core. The result is a difference in

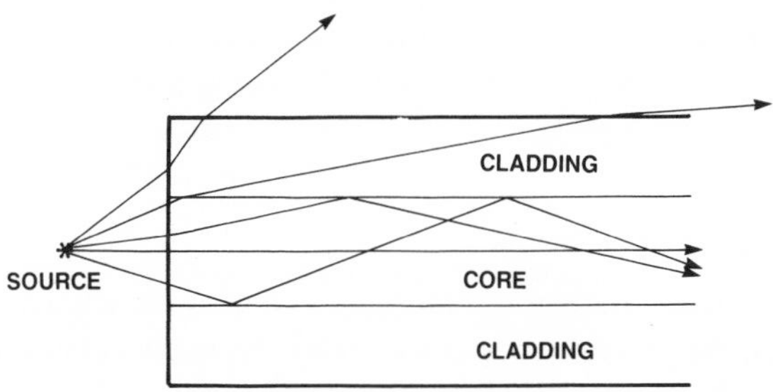

Figure 3.42. Transmission of optical rays along the core of an optical fiber.

arrival time at the far end which produces pulse spreading and can produce interference between successive pulses (intersymbol interference).

This dispersion can be corrected by shaping the refractive index of the core so that rays close to the central axis travel slower than rays at the edges of the core. This is the reason for using the graded index profile of example (b) in Figure 3.41. A significant improvement in signal bandwidth is achieved by this means. For step index fibers the product of distance and bandwidth may be some 20–30 MHz km; for graded index fibers, the product may be some 1–2 GHz km. Even better performance can be obtained from fibers with a very small core—for these *single-mode* fibers, the product may be 50–60 GHz km. The refractive index profile of a single mode fiber is shown in example (c) of Figure 3.41. Because the core diameter is comparable to the wavelength of the optical energy, the propagation mechanism can no longer be described in terms of rays but must be viewed as wave propagation governed by classical wave equations (Maxwell equations). While the majority of the energy is contained within the core, a significant (and essential) fraction travels in the cladding, so the cladding must be designed and controlled every bit as much as the core. This is in contrast to graded index fibers in which only the cladding–core interface and not the cladding *per se* influences performance. Launching a sensible amount of single-mode energy in cores only a few microns in diameter requires sophisticated technology. In contrast, the fat fiber profile illustrated in example (d) of Figure 3.41 is easy to fill and will transport a large number of propagating modes. For this reason, fat fibers are used for applications which require relatively low bandwidth signals.

The loss associated with an optical fiber is a function of wavelength. It is illustrated in Figure 3.43. Fiber making has reached such a level of perfection that measured values fall close to the theoretical limits imposed by photon scattering due to the mechanical scattering (Rayleigh scattering) and energy absorption (infrared absorption) properties of the lattice. The peaks in the measured curve are due to the presence of impurities consisting of various metallic and hydroxl (OH) ions. In the mid 1970s, the optimum wavelength for the operation of optical systems was around 0.85 μm and good quality optical fiber had losses of 3–4 dB/km. By 1980, the concentrated efforts of many researchers had improved the longer wavelength performance so that the optimum wavelength for the operation of optical systems was around 1.3 μm. At this wavelength, fiber losses were reduced to around 0.5 dB/km. Material dispersion is close to zero, making 1.3 μm graded index (multimode) operation possible with bandwidth–distance products of 2 GHz km, or more. At 1.55 μm, fiber loss is around 0.3 dB/km. This region is particularly attractive for single-mode operation since waveguide and material dispersions can be caused to balance out by the proper selection of basic materials and impurity profiles. New materials may make operation at longer wavelengths such as 4 μm possible. Work in this area has achieved performance equivalent to 0.85 μm fibers of the early 1970s.[18]

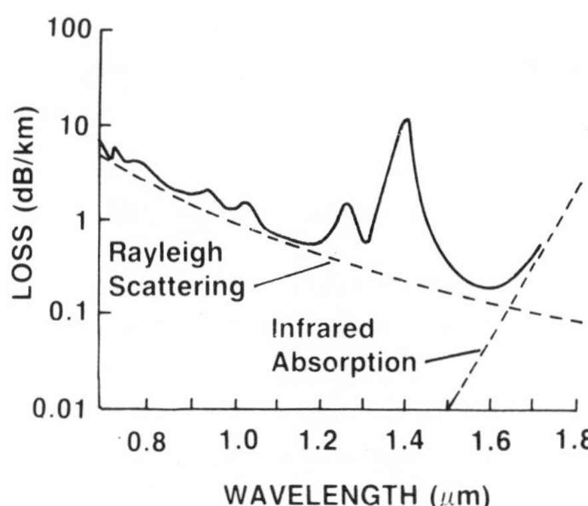

Figure 3.43. Loss associated with an optical fiber as a function of wavelength.

The absolute mechanical properties of optical fiber are virtually impossible to measure because glass is a brittle solid which cannot deform plastically to relieve local stresses. The strength of a given length of fiber is determined by the most severe flaw present. For a 1-μm flaw, it has been estimated that failure will occur at 84,000 pounds per square inch (psi). This is somewhat higher than the tensile strength of hard drawn copper wire (60,000–70,000 psi). Even though a well-designed cable is constructed so as to minimize residual stress, the fibers may be subject to forces that approach 20,000 psi due to bending during installation. During manufacture, it is usual to subject the fiber to a minimum proof test (50,000 psi, for instance).

3.12.2. Cables

Coated, proof-tested fibers are assembled into cables for application in telecommunication networks. The cable structure must protect each fiber against damage during manufacture, installation, and use. To some extent, making fiber cable is easier than making copper wire cable: the fiber is smaller, lighter, and more flexible. In addition, since glass is electromagnetically inert, it does not require shielding, nor is there a need to protect against induced voltages (caused by lightning strikes, for instance), so that many of the metal screens and grounding precautions included in wire cables can be left out. To some extent, fiber cable making is more difficult: the very smallness of the fiber makes the initial operations more delicate. At the end of manufacturing, the fiber must be free of stress because residual forces can cause small bends in the fiber which result in increased losses. Elimination of *microbending* becomes progressively more important as the intrinsic loss of the fiber is reduced. Using a combination of new and existing cabling techniques, fiber cables are now being manufactured which can be installed in a routine fashion by craftspersons. They are small, lightweight, and flexible and are made in longer lengths than copper wire cables because resistance to pulling through ducts, or over aerial pulleys, is greatly reduced. The longer length means fewer splice points.

3.12.3. Connections

The success with which fiber-to-fiber connections are made depends critically on the radial geometry of each fiber and on the preparation given to the fiber ends. A clean, right-angled cleave can be made by placing the fiber in light tension, bending it over a curved surface, and gently nicking the outside surface of the fiber with a sharp tool. Almost always the fiber will *cleave* squarely and cleanly. Tools are available for this purpose.

Radial geometry must be built into the fiber at the time of manufacture. The closer the circumferences of the fiber and core conform to perfect concentric circles and the more nearly the two fibers match, the better the connection that can be made. Core eccentricity and fiber and core ellipticity are quantities which must be minimized. They are illustrated in Figure 3.44. Further, tight tolerances

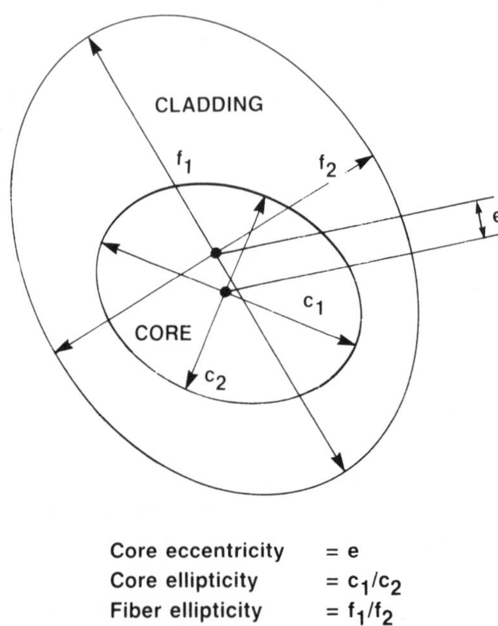

Figure 3.44. Radial geometry of the core and fiber.

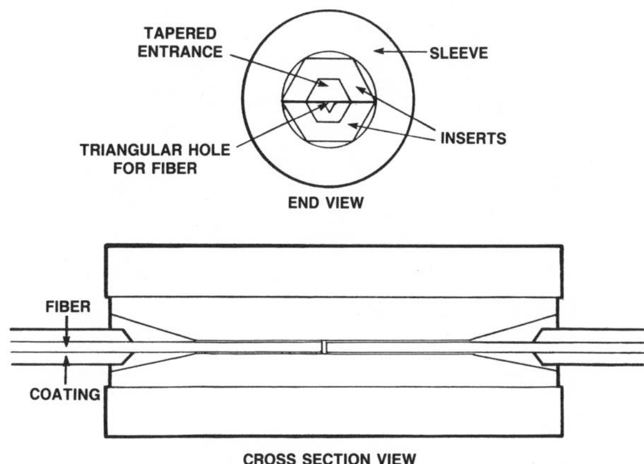

Figure 3.45. Drawing of an elastomeric tube splice.

must be placed on the diameters of the core and fiber so that any fiber can be joined to any other fiber without major mismatch and consequently high loss.

Permanent connections are called *splices*. They may be used to join fibers during cable making, or to join fibers as part of the process of connecting cables during installation. A number of approaches exist. They rely on aligning the axes of the cleaved fibers by reference to their outer surfaces. One popular technique employs a gaseous arc to fuse the fiber ends together. During fusion, surface tension effects between the molten ends produce some self-alignment. However, the heating may degrade the mechanical strength of the material around the join. Other techniques use accurately produced alignment members of metal, glass, ceramic, and plastic. An example of an elastomeric tube splice is shown in Figure 3.45. With well-made fiber, the use of this device can achieve under 0.1 dB loss per splice with multimode, 50-μm core, 125-μm outer diameter fibers, and around 0.1 dB loss per splice with single-mode, 7-μm core, 125-μm outer diameter fibers.

When semipermanent connections are required, such as in equipment bays and at points where the network must be rearranged frequently, screw-type connectors are used. By far the most successful employ lenses to enlarge and collimate the beam so that adequate alignment of the mating parts can be achieved with reasonable mechanical tolerances. An example of a lensed connector is given in Figure 3.46. Such devices introduce around 0.2 dB loss. It is important to recognize that optical fiber systems have advanced to parity with wire systems—the greatest concern of those engineering an installation is with the joints and connectors, not the medium itself.

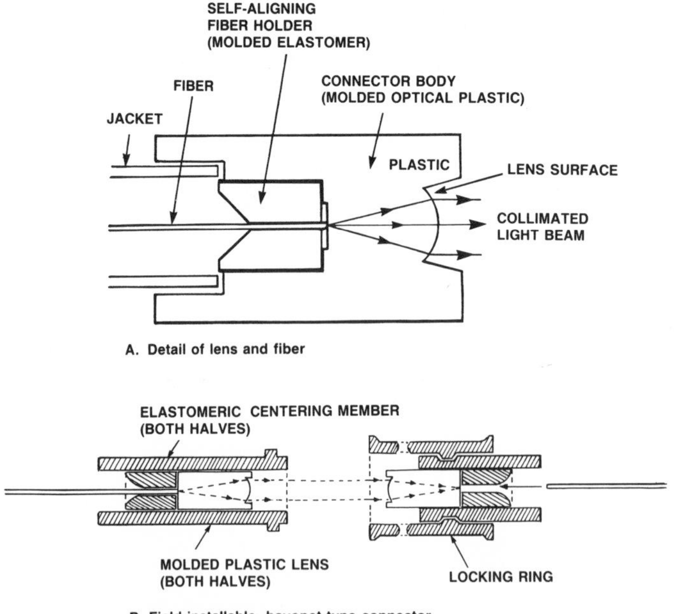

Figure 3.46. Drawing of a molded lens connector. (A) Detail of lens and fiber assembly; (B) field installable, bayonet-type connector.

3.12.4. Sources and Detectors

Optical energy is manifest in the form of *photons*. Members of the family of quanta, photons provide the binding force in electromagnetic interactions—they bind electrons to the nucleus. In some semiconductor materials, a voltage between the proper points on a specialized structure will cause the recombination of electrons and holes with the emission of photons and it is possible to modulate the light output by varying the applied voltage. In other structures, the absorption of a photon with sufficient energy will create an electron–hole pair and produce current flow.

Solid-state sources and detectors must

- operate at wavelengths which match the minima in the loss versus wavelength characteristics of optical fibers;
- be configured to launch adequate energy into the fiber; and
- have a lifetime which is compatible with other solid-state devices.

At 0.85 μm, GaAs-based structures are common, and at 1.3 and 1.55 μm, InP-based structures are used. Accelerated life testing predicts lifetimes of over 100 years. Two kinds of sources are employed: high-radiance light-emitting diodes, and semiconductor injection laser diodes.

The action of a light-emitting diode (LED) depends on

- the recombination of minority carriers at the p–n junction;
- the geometry of the device; and
- the physical arrangement of device and output fiber.

3.12.4.1. Construction

The construction of a typical high-radiance LED is shown in Figure 3.47. With the exception of the active or recombination layer, which is a small fraction of a micron thick, the other layers may be several microns thick. Within the *active region* in the InGaAsP layer, photons are created by the passage of current between the small contact at the base of the structure and the large area contact at the top of the structure. When the diode is forward-biased, minority carriers are injected into the active region, where they recombine, releasing photons. The *n*- and *p*-type cladding layers form a double-heterostructure which creates potential barriers preventing the minority carriers from diffusing out of the active region. Photons which are radiated in a direction included within the acceptance angle of the fiber become the useful light injected into an optical transmission system.

LEDs of the type shown in Figure 3.47 are *surface* emitters: they are known as Burrus diodes, after their inventor C.A. Burrus. An alternative structure known

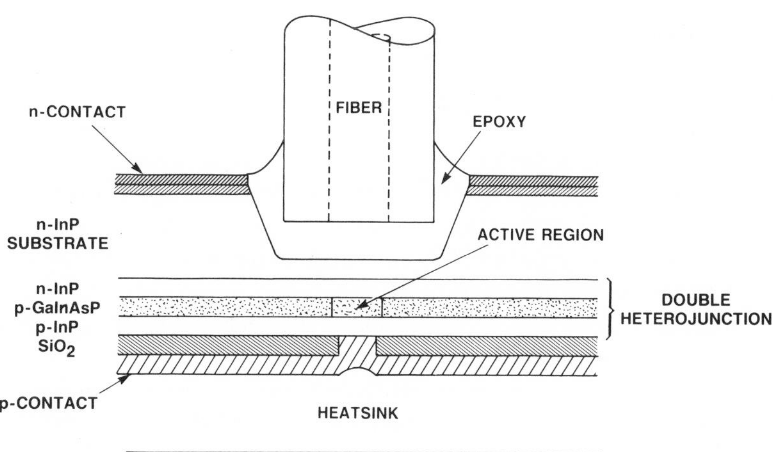

Figure 3.47. Structure of a high-radiance light-emitting diode.

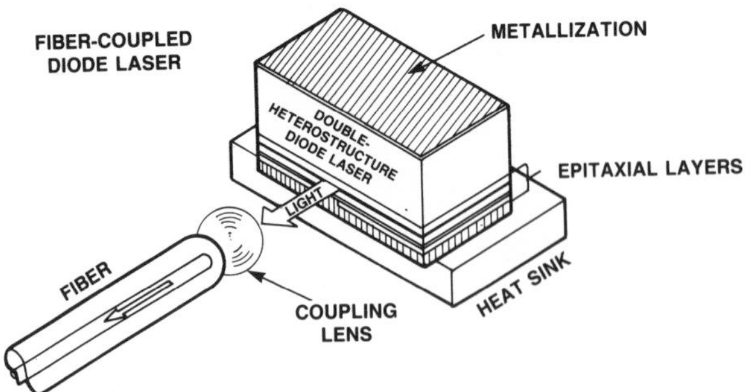

Figure 3.48. Structure of a double-heterostructure laser diode and its coupling to fiber.

as an *edge* emitter employs photons which radiate in directions which cross the current flow. In the same way that they are guided in optical fibers, photons striking the interface between the active layer and the cladding layers at angles greater than the critical angle are trapped within the active layer by the heterojunctions. As a consequence, the radiance of these devices can be higher than surface emitting devices.

Most semiconductor laser diodes (LDs) also employ a double heterojunction: a typical structure is shown in Figure 3.48. The ends of the laser are cleaved to produce mirrors which form an optical cavity (Fabry–Perot cavity). When the current through the recombination layer is sufficient to produce emission in excess of the cavity loss (internal and external), the optical feedback due to the mirrors rapidly stimulates further emission and causes lasing to occur. The result is a dramatic increase in radiance and a narrowing of the emission spectrum. Since the material in the cavity emits over a range of wavelengths, some are selectively enhanced by reflection at the end mirrors. The result is a strong central line (frequency or wavelength) surrounded by as many as a dozen other lines of smaller intensity which correspond to frequencies for which an integral number of wavelengths fit between the mirrors. They are known as longitudinal modes. The width of the active region determines the number of transverse modes which propagate. They can be controlled by the width of the stripe contact, by special processing to form heterojunction walls, or both techniques can be used together. A laser in which only one transverse mode propagates is known as a single-mode laser. A laser in which one transverse mode and one longitudinal mode propagates is a single-mode, single-frequency laser. Depending on the application, performance required, and wavelength, the optical aperture may vary from 2 to 10 μm wide by 0.1 to 0.2 μm thick. The optical power is radiated in a thin, wide-angle, fan beam. To ensure capture of as much energy as possible, a sphere or lens is often placed between the aperture and the end of the fiber to focus as much energy as possible into the acceptance cone.

At the other end of the transmission fiber, the problem is to receive and detect photons. The basic effect employed is the absorption of photon energy to produce an electron–hole pair which can be rapidly swept apart to produce a current. This effect may be enhanced by arranging that the electrons are raised to the conduction band in a relatively extensive region of high electric field strength (depletion region) so that they move rapidly and produce high displacement current. In a conventional diode made from *n*- and *p*-type material (PN diode), the depletion region may be shorter than the photon absorption region. This is corrected by introducing a region of undoped silicon (intrinsic silicon) between the doped materials to form a *p–i–n* diode (PIN diode).

If a very high reverse voltage is applied across a *p–n* junction, the primary photo current can be multiplied. This occurs because the primary carriers created by photon absorption are raised to a high enough energy that they are capable of creating fresh electron–hole pairs by impact ionization. Given even a modest multiplication, an avalanche is created which greatly amplifies the event. Devices in which this occurs are known as avalanche photodiodes (APDs). Typically, they may have gains of 100, and produce a better signal/noise ratio than a PIN detector.

Silicon devices (PIN diodes and APDs) are well suited to the detection of optical energy at wavelengths of 0.8–0.9 μm. Germanium devices are useful to 1.3 μm. At longer wavelengths (1.3 and 1.55 μm) materials similar to those used to construct longer-wavelength sources (InGaAs, for instance) are used.

3.12.4.2. Operating Characteristics

The light radiated into a fiber is a function of the physical construction of the source and the size of the fiber. Modern telecommunication systems employ 125 μm outside diameter fiber, and components are usually supplied with a fiber *pigtail* which can be connected to the transmission fiber using the techniques already described. Optical power output is measured at the end of the pigtail. Typically, for LEDs, it may be from 20 to 50 μW, and for LDs, it may be from 1 to 5 mW. The spectral width of the energy radiated may be 0.050 μm for LEDs and 0.001 μm for LDs. LEDs produce energy without coherence. LDs produce coherent energy. These differences make LDs the sources of choice for high-performance systems and limit LEDs to lower-speed, shorter-distance applications.

To achieve the best detection possible, all of the photons which emerge from the transmission fiber must impact on the active area of the detector. The area, then, must be some 100 μm in diameter. In fact, detectors are usually made with a reception area much larger than this so that mating the fiber to the detector does not call for a high degree of mechanical accuracy. The choice between PIN and APD is not as clear as LED versus LD. For silicon devices (employed at the shorter wavelengths) APDs have a firm performance advantage over PINs. For InGaAs and other devices (employed at the longer wavelengths) the performance of a PIN in combination with a low-noise FET amplifier is probably every bit as good as an APD.

3.12.5. Optical Multiplexing

In telecommunication systems, an important task is multiplexing—that is, the carriage of more than one optical carrier on the same fiber. In its simplest form an optical multiplexing device consists of wavelength-discriminating reflective filters as illustrated in Figure 3.49. In the first example (A), light at three wavelengths is amalgamated and transmitted in the same direction over a single fiber. Reflectors F_1 and F_4 pass light of wavelength λ_1 but reflect light of wavelengths λ_2 and λ_3: reflectors F_2 and F_3 pass λ_1 and λ_2 but reflect λ_3. In the second example (B), light at two wavelengths is passed in opposite directions over a single fiber. Reflectors F_5 and F_6 reflect λ_2 but pass λ_1. Multiplexers provide an easy way to increase the number of optical carriers on a single fiber: they are used to conserve fiber on long-distance links. In other applications, multiplexers are used to combine optical carriers which transport different messages (voice, data, or video, for instance) on a single fiber.

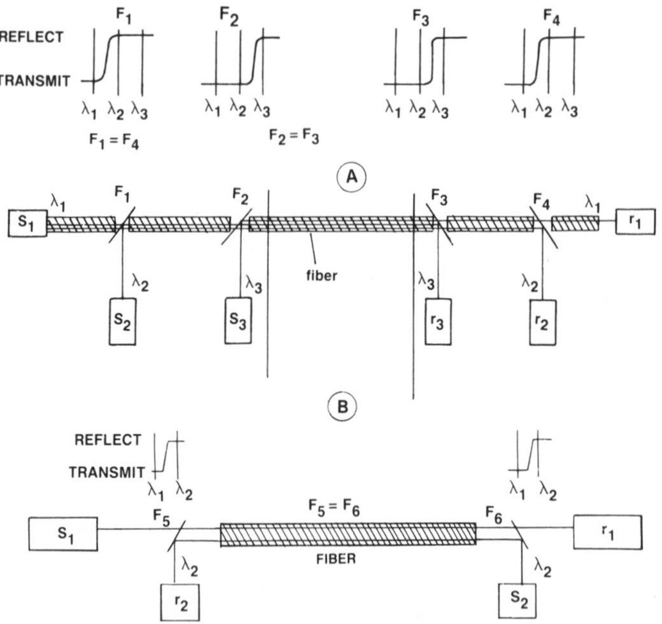

Figure 3.49. Examples of multiplexing light of several wavelengths over a single fiber using wavelength-selective reflective filters. s_n and r_n are a source and a receiver for light of wavelength λ_n.

3.12.6. An All-Optical System?

For optics to be more than a transmission medium—to have the potential of providing a complete, end-to-end, optical telecommunication channel—requires the ability to perform optical signal processing. At the very least, optical space or time division switching and optical repeating must be possible. While it is intriguing to consider that these functions may one day be implemented solely in the optical domain (no electronics), in the near term they will employ a combination of optics as the primary channel medium, and electronics as the primary control medium.

New compound semiconductors are being developed from materials with dissimilar crystal dimensions by growing extremely thin layers of one material on the other. Known as *superlattices,* the layers are one or two hundred angstroms (one or two hundredths of a micron, 1 or 2×10^{-8} m) thick. They are thin enough that the atoms in each succeeding layer can bond with the ones below through elastic distortion of the lattice. The layers are less than the electron mean free path so that electrons can generate photons without giving up energy to the crystal lattice (as vibrations). They may be the sort of devices which will make an all-optical system a 21st century possibility.

4

Facilities

Contemporary telecommunication facilities consist of a range of equipment which employs a spectrum of technologies. At its introduction, each type of equipment was a compromise between revolution and evolution as the designers sought to achieve greater reliability, lower cost, and new features under the constraint that it must connect to older equipment. The substantial investment in existing equipment has assured compatibility and forced the pattern of change to be that of controlled upgrade. In this chapter, we describe modern, primarily digital, equipment which provides the facilities supporting voice, data, and video telecommunication to business and residence customers. For convenience, the discussion is divided into three major sections: telephony, television and radio, and data communications.

A. TELEPHONY

In this section we discuss interexchange, exchange area, and customer premise equipment which support point-to-point voice traffic.

4.1. Interexchange Facilities

Interexchange facilities transport traffic between exchange areas, whether they be adjacent to each other or across the country, and transport messages on their way around the world. Interexchange facilities consist of transmission plant and specialized switching equipment connected in a range of networks.

4.1.1. Interexchange Transmission

Most long-distance traffic is carried on microwave radio, coaxial cable, and an increasing amount of optical fiber. A relatively small amount of traffic is also

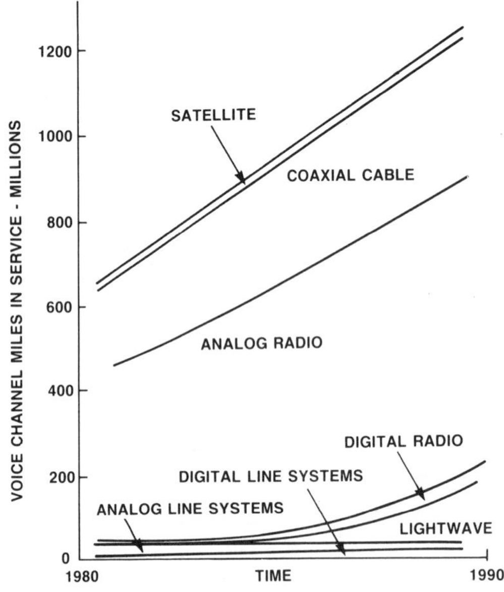

Figure 4.1. Estimate of total number of voice channel miles in service and the makeup of the types of facilities in use providing interexchange transmission in the 1980s in the United States.

carried by satellite circuits, as well as analog and digital line systems. An estimate of the number of channel miles provided by each system type is given in Figure 4.1. Discussions of analog and digital line systems will be reserved for Section 4.2.3.

4.1.1.1. Microwave Radio Systems

The externals of microwave radio have become a familiar sight following the installation of the first radio relay system between New York and Boston in 1947. By 1951, AT&T had extended it across the continent, and today the telltale steel or concrete towers, festooned with horn and dish antennas, are visible virtually everywhere in the developed world.

4.1.1.1.1. Analog Systems. More than two-thirds of all long-distance calls in the U.S. are carried over a network containing some 6000 microwave radio stations which operate in common carrier bands around 2, 4, 6, and 11 GHz and provide over 400 million circuit miles of transmission pathway. The modulation is largely frequency modulation (FM) and the input signal is a frequency division multiplex (FDM) configuration of single-sideband, suppressed carrier (SSB-SC), amplitude modulated (AM), 4-kHz telephone channels. In order to achieve a uniform signal-to-noise ratio in all channels, preemphasis is added to the baseband signal to compensate for the increase in system noise with frequency. This function is usually accomplished in the *entrance link* between the multiplexer and the radio system. Entrance links vary from a few feet to several miles. At the receiving end, a complementary amount of deemphasis restores the baseband signal to its original form. (Basic transmission concepts are contained in Section 3.5.)

The elements of an analog microwave radio system are shown in Figure 4.2. For simplicity, only one direction of transmission is shown. Four kilohertz voice channels are multiplexed to form a master group (or higher) baseband signal which is carried over the entrance link to the FM terminal and radio transmitter. The entrance link is a combination of interface equipment and a cable or optical fiber transmission medium which bridges the distance between the location of the multiplexer (usually a switching office) and the microwave radio equipment (usually at the antenna site). Besides providing preemphasis, electronic equipment associated with the entrance link amplifies the baseband signal, provides equalization to compensate for cable or fiber distortion and establishes the transmit level so as to be compatible with the other parts of the network. At the FM terminal, the baseband signal frequency modulates an intermediate frequency (IF) carrier (usually around 70 MHz) which is amplified before application to the radio transmitter. In this equipment, the IF carrier is converted to the assigned radio frequency (around 2, 4, 6, or 11 GHz), amplified

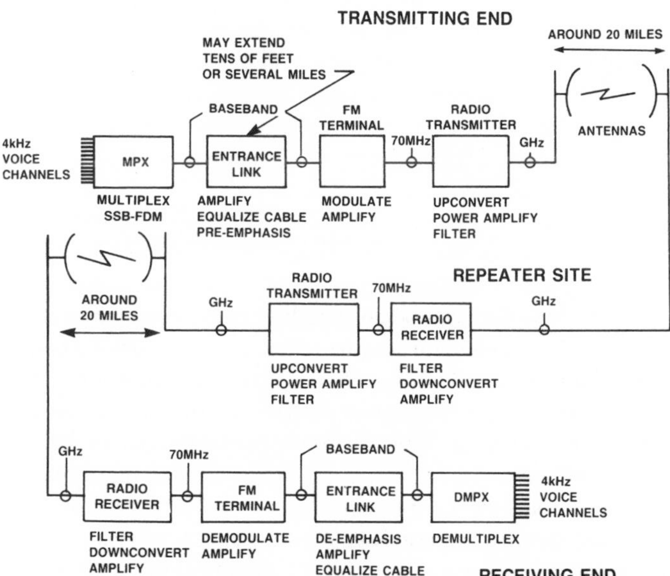

Figure 4.2. Basic components of an analog microwave radio link. Only one direction of transmission is shown. Actual systems employ several radio channels.

to a signal of around 1–10 W, passed through a filter to ensure the assigned bandwidth is not exceeded, and applied to the antenna to be radiated over a line-of-sight path to a repeater site. At the repeater, the spent signal is converted to IF, amplified, converted back to microwave frequency, and relayed on to the next site. At the receiving destination, the signal is decomposed to individual 4-kHz voice channels by reversing the processing used at the transmitting end. The reverse direction of transmission employs similar equipment, and most systems incorporate more than one radio channel, In a practical system, then, portions of the chain shown in Figure 4.2 are replicated several times (to provide additional radio channels) and the whole is duplicated (to provide transmission in the opposite direction). Other equipment provides spare radio channels, tests the performance of individual links, and provides for the rapid substitution of a spare channel when an active channel fails. Long-haul systems are designed to operate with up to 200 or more repeaters over distances of up to 4000 miles. Because the signal is passed from site to site, they are often referred to as microwave radio *relay* systems. A typical system (TH-3) can transport 14,400 two-way voice circuits using eight radio channels in each direction.

An increasing demand for long-haul circuits and the exhaustion of available frequency assignments in many areas has encouraged the substitution of a more efficient technique for the frequency modulation employed in the system described above. Since 1981, new microwave systems (AR6A) are being deployed which incorporate single sideband, suppressed carrier, amplitude modulation in the radio transmitter. A higher level of linear performance is maintained throughout the signal handling chain and special attention is given to correcting any frequency shifts due to demodulation and remodulation in the repeaters.[1] Six thousand voice channels can be fitted into the bandwidth previously used by 1,800 (in an FM system). The AR6A system can transport 48,000 two-way voice circuits using eight radio channels in each direction.

4.1.1.1.2. Digital Systems. A digital radio system is one in which the transmitted signal is modulated in such a way that it assumes a finite number of unique states defined by specific values of frequency, amplitude, or phase (or combinations of these parameters) so as to transport as much digital information as possible within the bandwidth allocated for radio operation (see Section 3.3.4). Most digital radios operate at 2, 6, 11, or 18 GHz. For intercity service, digital radio can be cheaper than analog radio: for transcontinental service, analog radio is cheaper than digital radio. This relation is brought about by the fact that the cost per channel mile of a digital transmission link is more than an analog transmission link, but that the cost per channel of digital terminal equipment is less than analog terminal equipment. With a radio employing 16 QAM, a 90 Mb/s stream can be transported in a bandwidth of 30 MHz. This is equivale. to 1344 voice channels. With SSB analog equipment, this number of channe would be carried in less than 6 MHz. Digital terminal equipment is cheaper pe voice channel because it is built from integrated circuit chips. If the signal is derived from a digital switch, the need for equipment which converts an analog channel to a digital voice channel and vice versa is eliminated and the need for additional multiplexing *may* be eliminated, making an even larger cost savings possible.

The basic components of a digital microwave system are shown in Figure 4.3. For the sake of simplicity, neither multiplexers nor entrance links are included in the diagram, and only one direction of transmission is illustrated. For specificity, the radio is assumed to employ 16 QAM modulation (see Section 3.3.4). Two binary digital information streams, which are derived from the same

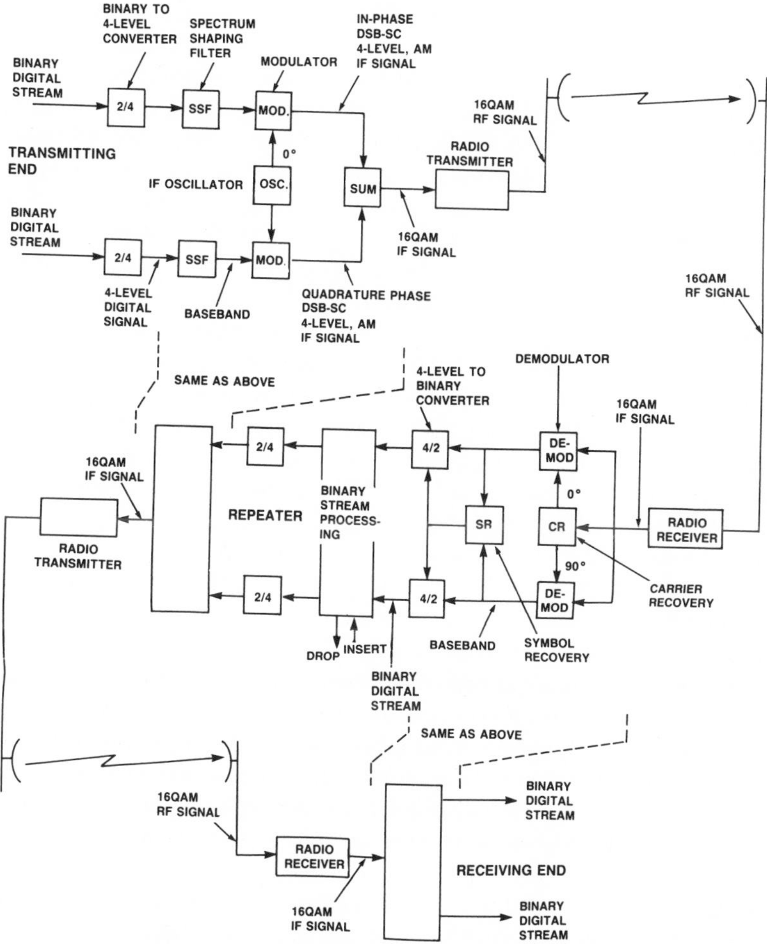

Figure 4.3. Basic components of a digital microwave system employing 16 QAM. Only one direction of transmission is shown. Actual systems employ several radio channels and may include diversity arrangments.

source, or are synchronized in some other way, enter at the top left-hand corner and are converted to four-level digital streams which employ raised cosine or other spectrum-minimizing pulses. After filtering to ensure that the signal spectrum falls within the desired passband, the four-level streams are used to amplitude modulate two intermediate frequency carriers of the same frequency, but whose phases differ by 90°. The result is two four-level double-sideband, suppressed carrier (DSB-SC), AM signals, one in-phase (I) with the local oscillator (IF) and the other at quadrature (Q) with it. Summing them produces a 16 QAM, IF signal, which is up-converted to the radio frequency (GHz), amplified, filtered to ensure that transmitted energy falls within authorized limits, and radiated from a suitable antenna. Because of the complexity of the signal, fading due to atmospheric effects has a much more deleterious effect on digital radio channel performance than on analog radio performance. A measure of compensation can be achieved by employing redundant RF paths in the expectation that while one may be degraded, the other will not. In actual systems, space, polarization, and frequency diversity are used with varying success (see Section 3.5.3.2).

In Section 3.3.4, it was pointed out that the more complex forms of digital modulation require greater values of signal-to-noise ratio for a given channel error rate. This is because the received, corrupted signal must be reconstructed automatically to obtain the information present in the discrete states which have been transmitted. To perform this processing accurately, the circuits require adequate decision margins so that they can distinguish between levels with the degree of certainty which matches the expectations of the users. An important parameter is the bit rate of the input streams. Without this, the recovery circuits may not make level decisions at the same point in successive bit periods, producing an increased potential for error. Worse, the recovery circuits may *lose synchronization* and generate an output stream which does not mimic the input stream, garbling the information, losing word and frame count, and generally rendering the channel useless. In a digital system, the repeater performs the function of amplifying the spent signal *and* restoring bit integrity, retiming and digital stream, and maintaining synchronization. It may also be the point at which some channels are separated from the channels which are carried through on the link so as to serve a facility in the vicinity of the repeater. Other channels may be added at the repeater point. This action is known as *drop and insert.*

All of these functions are incorporated in the repeater in Figure 4.3. The 16 QAM RF signal is received, amplified, converted to IF, and applied to phase demodulators. In addition, a separate carrier recovery circuit derives a signal which is equal in frequency and phase to the IF carrier suppressed at the transmitting end. In-phase and quadrature components are used to demodulate the signal to two, four-level baseband signals which are applied to four-level to binary converters. A symbol recovery circuit uses the signals to derive a control signal which matches the rate of the streams at the transmitting end. It is used to time the conversion process and ensure level sampling is done at the right instant. The reconstructed binary digital streams are processed to ensure word and frame integrity, parity, and other parameters. Channels to be dropped at this point are extracted, and channels to be inserted are incorporated. The outbound binary streams are reprocessed to form a 16 QAM RF signal which is relayed to the next repeater, or to the receiving end of the link where it is reconverted to the two original digital streams. The reverse direction of transmission employs similar equipment and most systems incorporate more than one radio channel. In a practical system, portions of the chain shown in Figure 4.3 are replicated several times (to provide additional radio channels) and the whole is duplicated (to provide transmission in the return direction). In addition, redundant equipment may be included to implement diversity in order to achieve reliable working in a nonbenign environment. A typical system (DR11-40) can transport 13,440 two-way voice circuits using ten radio channels in each direction.

4.1.1.2. Cable Systems

Coaxial cable transmission systems use FDM voice channels and SSB-SC modulation. The largest system carries 132,000 conversations over a 22-tube coaxial cable. Repeaters are spaced every mile and equalization must be applied to compensate for frequency- and temperature-dependent attenuation and noise characteristics. The loss of one mile of 3/8-in. coaxial cable is as low as 1 dB at low frequencies and as high as 30 dB at 60 MHz. The basic line repeater is designed to equalize this loss over a nominal repeater spacing. Where spans are used which are less than nominal length, line *build-out* networks are installed. These units provide the additional frequency-dependent loss to balance the fixed gain of the repeater. In long sections containing many repeaters, slight variations of gain in each repeater may add up to give an unacceptable frequency response after a number have been traversed. This can be compensated for by the inclusion of equalization inserted during installation and adjusted to match the characteristics of the link during testing. Yet one other effect is encountered. Cable loss characteristics vary with temperature. For cable buried at an average depth of 4 ft, the local temperature will vary approximately ±20°F over a year. This amounts to ±0.67 dB/mile at 60 MHz—an effect which rapidly dwarfs other variations. Automatic, temperature-compensating equalization is included in 10% to 20% of all repeaters, depending on terrain and environment. Often, they are known as *regulators* or regulating repeaters. The high level of engineering required by coaxial cable systems is expensive, and they can be costly to maintain. The availability of optical fiber systems with superior properties has probably stopped future new installations of coaxial cable systems.

4.1.1.3. Optical Fiber Systems

In 1977, the first optical fiber transmission systems were placed in trial service in the United States. Since then, progress has been swift, and already *second* generation systems are entering service. The promise of high-capacity,

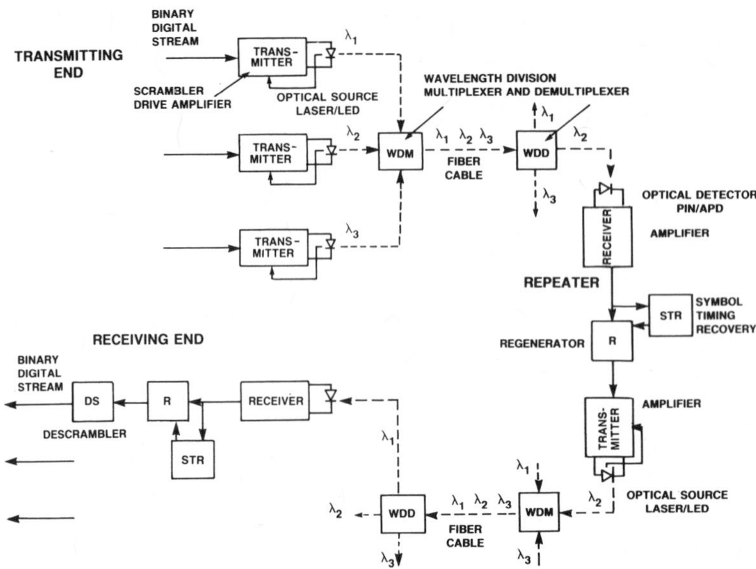

Figure 4.4. Basic components of an optical fiber transmission system which incorporates wavelength division multiplexing. In practice, a system will consist of many fibers.

interference free, easy to use light wave systems has been substantially realized. Figure 4.4 shows the basic components of an optical fiber system which incorporates wavelength division multiplexing (see Section 3.12.5). Three separate, independent digital streams are applied to three transmitters which drive optical sources generating energy at three wavelengths. They are carried on separate fibers to a multiplexing device which puts all three into the same fiber. The three optical signals are transported for many kilometers to the first repeater site. Here, they are demultiplexed and each is converted to an electronic signal which is received, regenerated, and amplified before being converted back into optical energy, remultiplexed, and relayed to the next repeater, or the receiving end of the link. The reverse direction of transmission employs similar equipment, and many systems use multiple fibers. In a practical system, portions of the chain shown in Figure 4.4 are replicated several times (to serve additional fibers) and the whole is duplicated (to provide transmission in the return direction).

A system of the type described (FT3C) can transport 240,000 two-way voice circuits on a 144-fiber cable. This is almost twice the voice circuit capacity of the largest coaxial cable system. In addition, since the optical properties of fiber are unaffected by ambient temperature, regulators are not required, and the repeater span is almost ten times the coaxial cable repeater spacing. Based on an earlier system (FT3) which is intended for backbone routing in metropolitan areas,[2] the FT3C system employs wavelength division multiplexing to support three optical carriers modulated at 90 Mb/s in graded index, multimode fiber. A higher bit rate of 432 Mb/s is planned for the FT4E-432 system (to be introduced in 1985) which will employ a single long-wavelength (1.3 μm) optical carrier in singlemode fiber. The average repeater spacing will be 32 km (20 miles). (Optical fiber, cables, sources, detectors, and other devices are discussed in Section 3.12.)

4.1.1.4. Satellite Communications

Since 1957, when the launching of Sputnik 1 from Tyura-Tam in Soviet Central Asia opened the space age, several thousand satellites have orbited the earth. Probably less than 10% of this total were used for civilian communications. Modern communications satellites are placed in orbit some 36,000 km (22,300 miles) above the equator, a point where the gravitational attraction of earth is equalized by the centrifugal force created by rotating about the earth's axis at the same angular velocity as the earth rotates. In this *geostationary* orbit, the satellite appears stationary from earth and can illuminate approximately one-third of the earth's surface.

To avoid signal interference between satellites, both space and frequency diversity are employed. In that portion of the geostationary orbit covering the United States (approximately 90° to 130° West longitude provides 50 states coverage: in addition 60° to 90° West longitude provides 48 states coverage) satellites are spaced at intervals of 2° (or greater). Satellites employ frequencies in C band (around 4 GHz, space to earth, and 6 GHz, earth to space) and Ku band (around 11 and 12 GHz, space to earth, and 14 GHz, earth to space). Additional frequencies are allocated in Ka band (around 18 GHz, space to earth, and 30 GHz earth to space). At 4/6 GHz, interference with terrestrial microwave networks which share the same frequency bands may restrict opportunities for siting earth stations. At the higher frequencies (Ku and Ka bands), the absence of terrestrial facilities competing for the same frequencies makes the selection of sites for earth stations easier. However, atmospheric attenuation due to rainfall, ice crystals, and fog can be significant. During heavy rainstorms in the eastern and southern regions of the U.S., diversity stations may be required to assure reliable signal reception.

In the simplest case, a satellite in orbit is a repeater in the sky. The on-board electronics capture the signals from earth, convert them to a lower frequency, and transmit them back to earth. A message originating from a single point within the field of view of the satellite can be relayed to any point on the one-third of the earth's surface illuminated by the satellite. The relayed message can be broadcast to the entire area, in which case it can be intercepted by everyone, or dispatched to a specific area through the use of a spot beam antenna, in which case only those within the area illuminated can receive it. In either case, there will be a delay of 0.25 to 0.3 s between transmission and reception back on earth due to the distance the energy must travel. The variation in delay reflects the difference in total path length between points immediately under the satellite, and points on the edge of the illuminated area.

The evolution of satellite communications from terrestrial microwave technology has resulted in the development of systems employing both analog and digital multiplexing and modulation techniques. In addition, new techniques have been invented to take advantage of the special features of the medium. Many satellites operating in C band use 24 transponders over the 500-MHz bandwidth available to them. Each transponder has a passband of 34 MHz, is separated by 6 MHz from its neighbors, and carries 1,200 one-way analog voice channels using FDM/FM techniques. With companded single-sideband modulation, this capacity can be increased to close to 3,000 channels. Twelve transponders receive and transmit horizontally polarized signals, and 12 transponders receive and transmit vertically polarized signals. The center frequencies of horizontally and vertically polarized transponders are offset by 20 MHz so that the peak power in each occurs in the guard band separating the two transponders of quadrature polarization which share the channel. The principle of a repeater-in-the-sky satellite operating in C band and employing 24 transponders using horizontal and vertical polarization in both the receive and transmit bands is shown in Figure 4.5. Only one direction of transmission and one radio channel are shown. Not all satellites use the frequency plan illustrated. For instance, transponders with 50, 72, and 240 MHz bandwidths are in use and other bandwidths are used at Ku and will be used at Ka frequencies. In Ku band, as many as 86,400 two-way voice channels will be available over a single satellite. A theoretical study shows an ultimate capacity of about 30 Gb/s, or over 200,000 conversations.(3)

frequency used and the distance traversed. For stations directly under the satellite, the loss on each link will vary from around 200 dB at C band to 300 dB at Ka band. For stations on the edge of the illuminated area, the loss on each link will vary from around 300 dB at C band to 450 dB at Ka band. Since the satellite transmitter power is supplied from the on-board power supply which is charged from solar cells, it is important that it be minimized. For this reason, the down-link frequency is always less than the up-link frequency. Another way of saving down-link power is to employ shaped-beam antennas which concentrate the radiated power on specific areas of the global surface. Thus, satellites intended to carry traffic between continents separated by an ocean employ antennas which concentrate the radiated energy on the landmasses, not the ocean. Satellites intended for domestic communications within the United States illuminate only the 48-state continental landmass with additional spot beams focused on Alaska, Hawaii, and Puerto Rico. In fact, even in the 48 states, special beams may be employed to serve the major centers of population.

4.1.1.4.1. Channel Access Schemes. The large area coverage and multiple

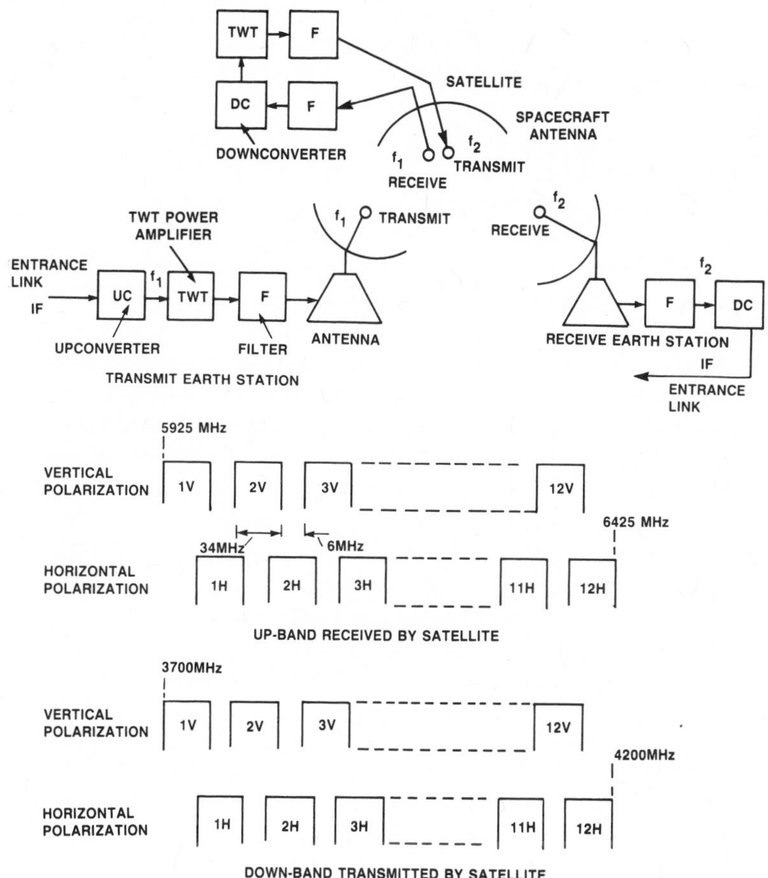

Figure 4.5. Principle of repeater-in-the-sky satellite operating in C band and employing horizontal and vertical polarization in both the receive and transmit bands.

connectivity that a satellite provides require that access must be carefully controlled so that a large number of earth stations can be interconnected through the same satellite simultaneously. Channels may be preassigned for routes with heavy or continuous traffic which require dedicated circuits, such as high-level connections in a national telephone network. Channels may also be assigned on the basis of demand. Called DAMA (demand-assigned multiple access), this technique can be implemented in the frequency or time domain.

Frequency-Division Multiple Access (FDMA). In FDMA each earth station is assigned a transponder, or portion of a transponder, and transmits a multiplexed signal within these frequency limits. The earth station receiving one or more of the messages replies on its own assigned channel. While each earth station only transmits within its assigned band(s), it must have a capability which covers all of the frequencies assigned to the locations with which it is to communicate. Taken one step further, the frequency bands can be assigned dynamically, as required, by a central control facility. The calling earth station communicates to the control station requesting circuits for another station. Control selects a frequency pair and notifies both stations of the assignment. This technique has been used with manual and computer controlled stations to provide single voice channel connections in sparsely populated areas of the world. Called SCPC-DAMA (single channel per carrier DAMA) it requires each earth station to be able to transmit and receive on several frequencies. FDMA techniques result in the simultaneous presence of several modulated carriers in a transponder. To prevent severe cross-modulation caused by nonlinearities in the amplifier, the power on each carrier must be reduced (or *backed-off*) from the level it could have if operating with only one carrier. Close coordination of up-link power levels is required to obtain maximum performance from each transponder.

Time-Division Multiple Access (TDMA). In TDMA, unique time slots are assigned during which each earth station has full and exclusive use of the available bandwidth. Using moderately efficient digital modulation schemes (4 PSK, for example), bit rates of 100 Mb/s are used with a transponder of 80 MHz bandwidth, and 600 Mb/s are used with a transponder of 500 MHz bandwidth, the entire spectrum available to a Ku band satellite. The principle of a TDMA system is shown in Figure 4.6. Messages are buffered at the earth stations and sent in

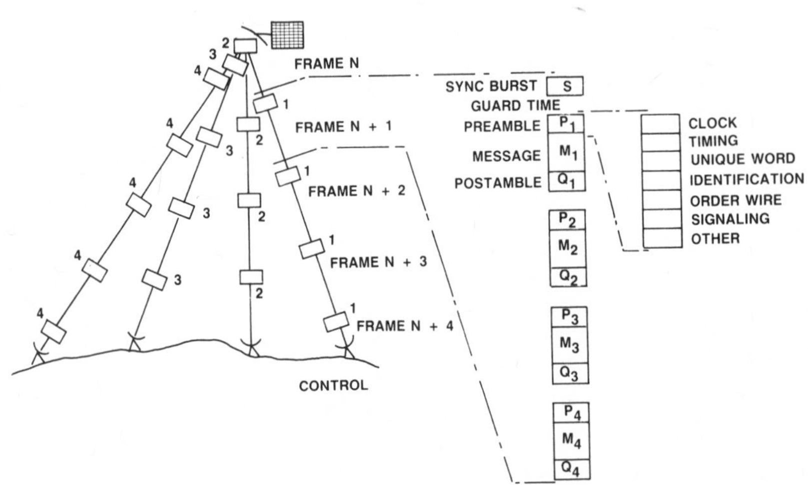

Figure 4.6. TDMA operation.

bursts so timed that they arrive to be relayed back to earth in a sequence which allows the satellite to serve each of the earth stations in turn. In the example, bursts are shown transiting the link in an organized sequence. Because of the long path, several discrete message blocks exist between each station and the satellite.

A set of message blocks (one from each station) constitutes a *frame*. It starts with a synchronizing signal (synch burst) from the control station which includes a unique word for easy recognition. To determine when to send their message bursts, all other stations use this signal as a reference. Transmit time is computed from the length of time it takes to receive their own messages back, and the sequence assigned to them by control. Each message is opened by a preamble block which contains block and timing information, station identification, and other data and may be closed by a block which performs parity checking and other functions, and ends the sequence. While it is being relayed by the satellite, each burst uses the full power available. There is no intermodulation distortion as in FDMA.

4.1.1.4.2. Spot Beam and Satellite Switching. Restricting the radiated energy to small areas of the field of view surrounding the terrestrial receiver allows better use of satellite power, provides some privacy by limiting the destination area, and permits the available radiofrequency spectrum to be used over and over again. In each spot beam, the entire assigned bandwidth can be available for use provided that adequate levels of isolation are maintained between the beams. This can be accomplished by designing an antenna with a radiation footprint which is a maximum in its intended service area and a minimum in areas served by other spot beams. To the degree to which this cannot be accomplished, it restricts the unlimited reuse of the frequency band.

The field of view of a geostationary satellite system spans some ten time zones. During the day, the pattern of traffic flow changes as peak activity moves westward following the sun. To make maximum use of the communication capacity of the satellite, switches can be used to change connections between transponders and fixed antennas to better serve the traffic peaks. In the simplest case, they reassign transponders to pairs of spot beams at the direction of the network control center in response to slowly changing traffic patterns. In the most ambitious system switching may be done within a frame on a message-by-message basis. The principle of SS-TDMA (satellite-switched TDMA) is shown in Figure 4.7.

4.1.1.4.3. Delay and Echo. For a telephone conversation completed over a satellite there is a one-way (one-hop) delay of 0.25–0.3 seconds, and a two-way (two-hop) delay of 0.5–0.6 seconds. For talkers engaged in a spirited conversation in which they interrupt each other aggressively, these delays can be a serious impediment which effectively disrupts communication. To overcome it, the speakers must pause between sentences, wait longer than usual for a response, and control their desire to interrupt each other. Under these circumstances communication is possible. However, pure delay is very rarely encountered. Invariably, it is accompanied by the reflection of some of the speaker's sound which returns as an echo that can severely degrade the channel. Reflections are generated at any point in the talking circuit at which the signal encounters a change in impedance. Their effect depends on their amplitude, spectral content, and circuit delay. If the time delay is short, the echo is not objectionable—in fact, it adds a quality known as reverberation, which is preferred by many people. If the time delay exceeds many milliseconds, the echo becomes objectionable, and must be attenuated. In fact, the longer the delay, the greater the attenuation must be in order to achieve an acceptable conversation.[4]

The major source of reflections in telephony is the point at which four-wire circuits become two-wire circuits. In a connection completed through digital end offices, this occurs at the hybrid transformer located on the line card which interfaces the two-wire customer loop to the separate go and return circuits (four-wire equivalent) in the switch. A lumped network consisting of a 900-Ω resistor and a 2-μF capacitor is used to balance the transformer bridge. Because customer loops vary widely in length and condition, this network provides an exact balance for only a few loops. Almost all hybrid transformers produce an attenuated, distorted version of the input speech which is reflected back to the originator.

The location of the hybrid transformers in the switches providing access to a satellite link is shown in Figure 4.8. Two techniques are used to prevent the reflection of the speaker's voice returning to the transmitting end of the con-

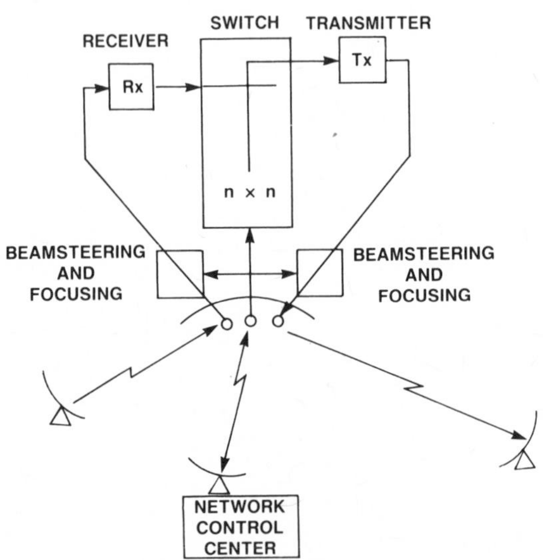

Figure 4.7. The principle of satellite-switched TDMA (SS-TDMA).

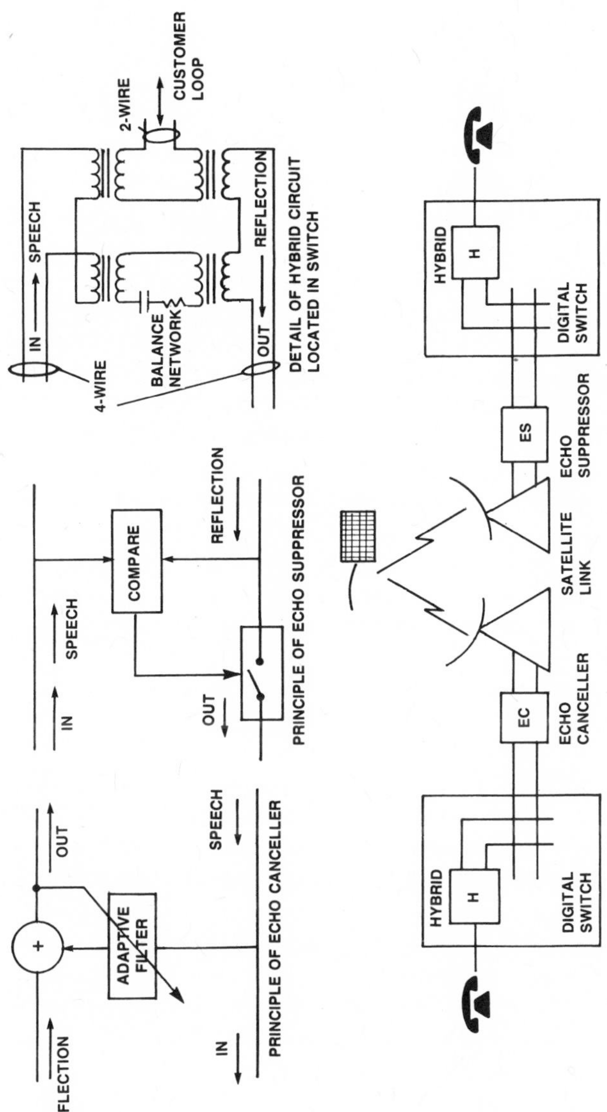

Figure 4.8. Location of hybrids, an echo suppressor, and an echo canceller in a satellite connection.

nection. One, echo suppression, uses circuits to open a switch in the return circuit when a speech signal is detected in the in-bound path. The other, echo cancellation, uses a digital circuit which adapts to the input conditions to automatically synthesize a replica of the echo path, forms an estimate of the reflection, and subtracts this signal from the reflection on the out-bound path, cancelling it. Of the two, adaptive cancellation is preferred since it leaves both directions of transmission intact. Suppression prevents echos—it also prevents interruption until the speaker has finished. This is an objectionable feature for some applications. Both the suppressor and canceller are shown associated with the earth station complexes. They are also associated with switching complexes since echo control is usually applied to all terrestrial calls between points over 2000 miles apart.

4.1.1.5. Submarine Cables

The first transoceanic submarine cable for telephone use was placed in service across the Atlantic Ocean in 1956 as a joint project of AT&T and the telephone administrations of Canada and the United Kingdom. Known as TAT-1 it was retired in 1978, after 22 years of service. The two coaxial cables provided thirty-six 4 kHz channels, a capacity which was later expanded to forty-eight 3 kHz channels. The first transistorized transatlantic system was placed in service in 1968. At the present time, well over 200 submarine cable links are operational around the world. The most modern provide some 4,000 two-way voice channels per coaxial cable and use single-sideband AM techniques similar to land-based systems.

4.1.1.5.1. Fiber Optics Applications. As in transcontinental applications, the maturing of single-mode, optical fiber technology has restricted further development of coaxial cable systems. Before the end of this decade, transoceanic cables will be in place which carry tens of thousands of voice circuits over multiple fiber cable. Because of the opportunity to employ integrated electronics, the systems will be digital, and will have repeaters spaced by 20 miles or more.

4.1.1.5.2. Time Assignment Speech Interpolation (TASI). Over the years, the channel capacities of in-place submarine cable systems have been enhanced by electronic techniques for circuit multiplication. They take advantage of the fact that, in an average conversation, each party speaks for no more than 40% of the time. Thus, each one-way connection is unused 60% of the time. By using sensitive speech detectors and fast switches, TASI assigns channels to speakers only at the time they are speaking. If the channels are kept filled, approximately twice as many conversations can be carried over a given number of circuits. This gain is known as *TASI advantage*. TASI introduces significant delay: time is required to detect the presence of a speech burst, select an available channel, connect the speech to it and transmit the information to the receiving TASI terminal where the voice channels are reconstructed. All together the delay amounts to some 50 ms. This is more than enough to require some form of echo control.

4.1.1.6. *Speech Interpolation*

The original TASI equipment was implemented by and installed in analog facilities. It was expensive and could only be used economically in conjunction with costly equipment such as submarine cables. The development of digital signal-processing techniques now makes it possible to apply the principle to analog voice connections between points separated by only a few hundred miles. The principle of a digitally implemented speech interpolation facility used to multiplex $2n$ analog voice circuits on an analog line of approximately n channels is shown in Figure 4.9. Voice signals are converted to digital signals at the user interface, the presence of speech is detected, and this information and channel number are provided to the control. Meanwhile, the speech is stored. Control finds a free channel, signals the receiving end giving the original channel number and transmission channel number, assembles the talkspurt, and dispatches it. Return signals are received and connected to the proper output channel under

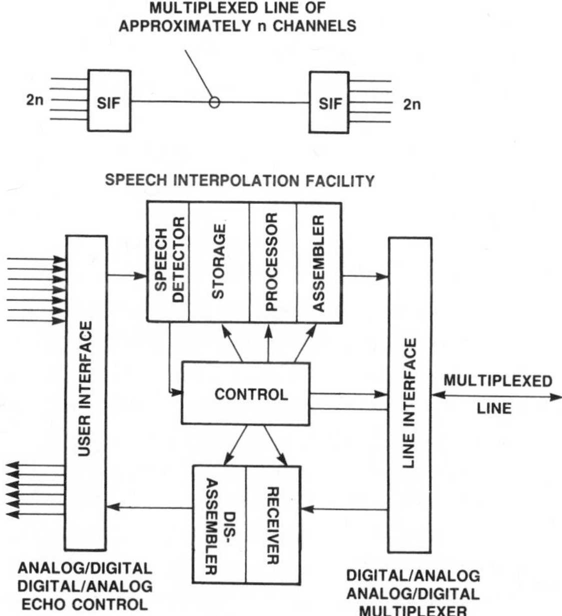

Figure 4.9. The principle of a speech interpolation facility (SIF) used to multiplex $2n$ analog voice circuits on a line of approximately n channels.

the direction of control. Digital-to-analog conversion and echo control are performed in the user interface unit.

Processing may be included in the sending channel to provide some measure of graceful reduction in quality as the system overloads. With analog voice, the strategies are limited. At the first level, some variation of storage delay is accepted as the incoming burst must wait for a free channel. At the second level, the front end of the burst is clipped so as to transmit the remainder of the burst within acceptable delay limits. At the third, and final level, the system may block all incoming calls for a few milliseconds to allow the channels to clear.

With digital voice and digital transmission systems, a much wider range of processing strategies is available. Bits may be dropped, reducing the digital voice samples to seven, six, or five bits momentarily; variable sampling rate delta modulation can be employed; so can adaptive differential PCM (see Section 3.2.2.2.1). These techniques form the basis for what is known as *digital speech interpolation* (DSI). They are able to compress voice traffic into less than one-half the normal channels and to handle traffic overloads gracefully.

4.1.1.7. Transmultiplexers

With the penetration of digital equipment (PCM) into what was an exclusively analog world, and the dominance of analog transmission (FDM) systems, FDM to PCM and PCM to FDM conversions are required at many points in the network. Two applications are illustrated in Figure 4.10. One shows transmultiplexers (TMUXs) providing PCM to FDM and FDM to PCM conversions so that an analog multiplexed transmission medium can be employed between the digital ports of two digital switches. The other illustrates an analog multiplexed transmission medium connected to a digital satellite. Complicated digital signal processing techniques are used to effect the transformation which depends on the ability to perform digital single-sideband modulation and demodulation. Modulation is achieved through an operation known as digital sample rate *interpolation,* which is the creation of a sequence of samples at a *higher* rate than an original sequence. Demodulation is achieved through an operation known as digital sample rate *decimation,* which is the creation of a sequence of samples at a *lower* rate than an original sequence. For a TMUX which handles 24 channels, the rates are noted in Figure 4.13. In addition to handling the voice channels, the equipment must also convert signaling from one system to the other, and make special provisions for data, which may be present at 56 kb/s in the PCM stream but cannot be forwarded over FDM facilities. For purposes of illustration, a 24-channel TMUX was chosen. This is the lowest level at which translation can take place. Other interfaces are possible, such as two FDM basic supergroups (2 × 60 channels) to five PCM T1 systems. The significant computing power required to implement either system is available in today's ICs, and entire TMUXs will be reduced to single chips in the future.[5]

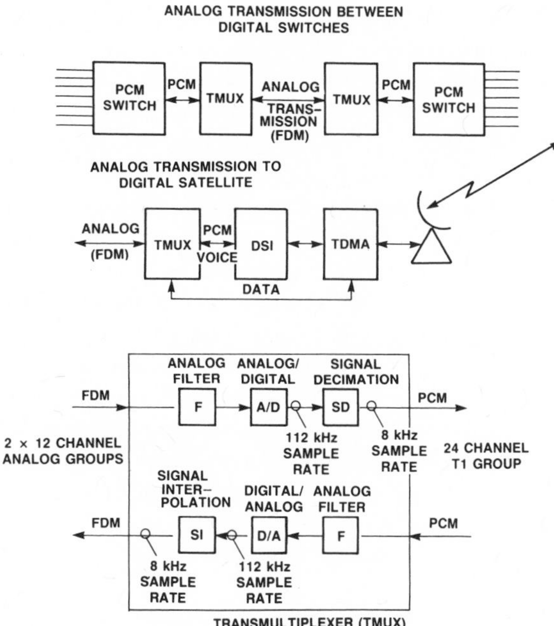

Figure 4.10. Principle and applications of a transmultiplexer.

4.1.2. Interexchange Switching

Transmission facilities are connected into a network by switches. In this way, messages are concentrated so that capacity is shared and alternative routes are provided, making it possible to accommodate surges of traffic—or survive physical disasters. The links are connected by switches which serve as the nodes of the network.

4.1.2.1. Network Architecture

From a functional standpoint, a network carries messages from one point to another on demand. From a physical standpoint the network consists of nodes (switches), links (transmission systems), and stations (terminals). In the abstract, the network is a stochastic service system which handles streams of telephone traffic made up of random events signifying the arrival of calls requesting connections from one place to another. Usually, the traffic is offered to the particular set of links directly serving the requested destination. If they are busy, the traffic *overflows* to an *alternative* route formed by other links which is generally longer than the direct route. The arrangements which cause these connections to be selected are usually called *homing* arrangements and are defined in the network routing plan. Most networks consist of a hierarchy of nodes, each of which receives the traffic in sequence until a free alternative route is found. If the call does not find a free link on the direct route, or on any alternative routes, the call cannot proceed, and the caller receives an *all-trunks-busy* signal.

The configuration of an economical, survivable interexchange network depends on the facilities available and the pattern of customer demand. Approximately 93% of all interexchange voice traffic in the United States is carried by the network formed from the long-distance facilities previously owned by the Bell Operating Companies, and the Long Lines Department of AT&T (see Sections 1.4.3. and 5.5.1.1). It is operated by AT&T Communications Corporation and may be referred to as the ATTIX (AT&T interexchange) network. More than 80% of the interexchange circuits in the ATTIX network are served by stored program controlled (SPC) electronic switching systems (ESS), and almost 90% of these circuits are supervised by common channel, interoffice signaling (CCIS). Over time, the in-place facilities evolved into a multilevel, hierarchical structure called, appropriately, the SPC network. Evolution is continuing: in some areas the network is being changed to a single level of fully connected ESSs to form a dynamic, nonhierarchical routing (DNHR) network.

4.1.2.1.1. Stored Program Controlled (SPC) Network. The SPC network is that set of SPC ESSs which are interconnected by CCIS. Common channel signaling is a natural accompaniment to computer control: over a separate link, data are sent directly from the computer controlling one switch to the computer controlling another to establish connections, or to exchange network management information, or for other purposes. By separating signaling from the speech path, it is not necessary to employ the same route, and the signaling channels have come to be star connected networks interconnect*ing* the ESSs and interconnect*ed* by signal transfer points (STPs). Since each data link may handle information for a thousand, or more, voice channels simultaneously, it is customary to provide redundant configurations to ensure physical security.

The major elements of the SPC network are shown in Figure 4.11. The interexchange network consists of four levels of switching centers which are interconnected both vertically and horizontally, to provide an efficient set of connecting links over which customer C1 may communicate with customer C2. Depending on the locations of the customers, the specific transmission links available, and the amount of other traffic in the network, they may be connected directly through TC1, i.e., EO1 → AT1 → TC1 → AT2 → EO2, through TC2, if the connection AT1 → TC2 exists, through TC1 → TC2, or any other combination available up to TC1 → PC1 → SC1 → RC1 → RC2 → SC2 → PC2 → TC2. The exact routing will have been derived from control messages passed over the CCIS network. Each STP can communicate with a network control point (NCP) for access to information to support nationwide services such as expanded 800-calling and automated calling-card services.

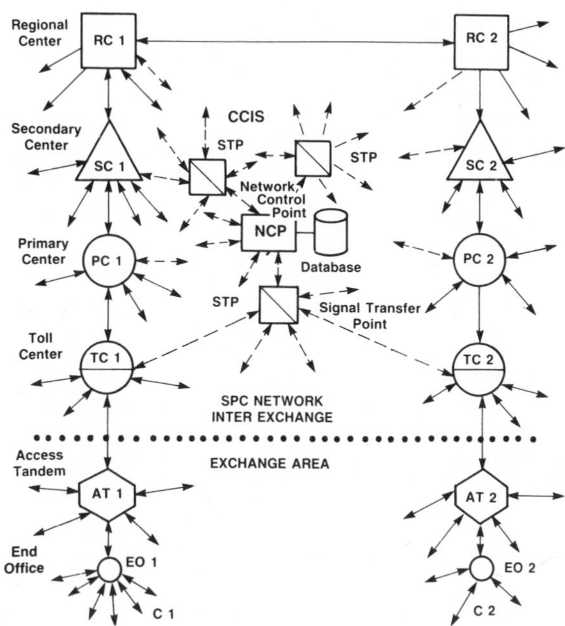

Figure 4.11. Representation of SPC network including CCIS and connection to exchange area access tandem.

4.1.2.1.2. Dynamic Nonhierarchical Routing (DNHR) Network. The four-level structure of the SPC network was set in place long before the first ESS was developed. Integrated circuit technology has led to the prospect of larger and larger switches and more powerful processors. Where traffic warrants, the result is that a multiply connected single level of large digital switches can economically handle all of the connections made by the four levels shown in Figure 4.11. In smaller metropolitan and rural areas, where traffic is insufficient to support direct connections to these switches, hierarchical switching levels will be used up to the level at which DNHR can be supported. In Figure 4.12, two customers are shown connected through a partial hierarchical network (SPC network) and, where traffic is intense enough, a DNHR network. Dynamic routing rules are applied so as to take account of time-of-day as the busy hour moves westward, and to ensure that any message does not transit more than two links in the DNHR network. Further, the total link length will be controlled so that it does not exceed twice the length of the direct path, plus 100 miles, and the maximum distance without echo control will be 2400 miles.

4.1.2.1.3. Network Control. Real time monitoring and control of the net-

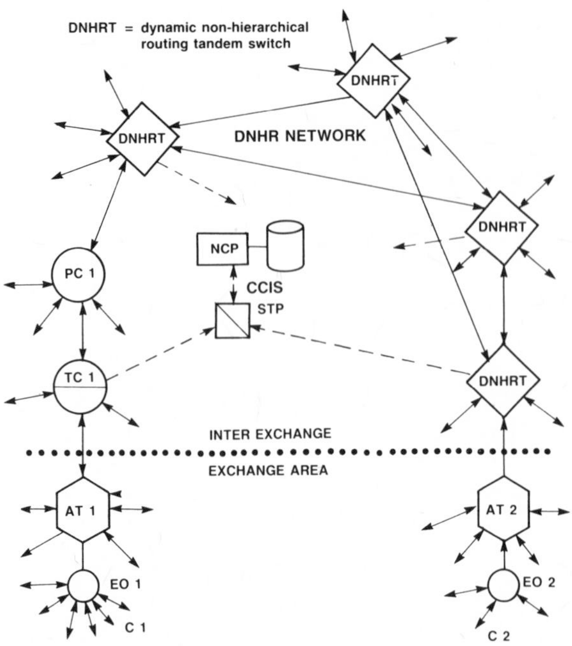

Figure 4.12. Representation of DNHR network including CCIS, partial hierarchical network, and connection to exchange area access tandem.

work is achieved through a coordinated set of operations centers (OCs). At present, there are

- 27 network management centers (NMCs);
- 10 regional operations centers which coordinate NMC activities within each of ten switching regions in the United States, and;
- a national network operations center which coordinates the activities of these centers and two Canadian regional management centers.

They work together to assure that optimum call-carrying capacity is maintained despite equipment failures, natural disasters, and unusual calling patterns. To accomplish this, each OC is the focal point for information concerning the real-time performance of the portion of the network under its management. This information consists of short-term average traffic data and frequent network status data for selected switching systems and transmission facilities. It is used to distribute traffic loads among links and nodes to meet customer demand. In real time, this is done by passing instructions to nodes to modify traffic routing patterns (homing arrangements) so as to use alternative routes which have capacity available. In the longer term, the information is used to plan new nodes and transmission routes or to expand existing facilities. The separation of exchange area and interexchange facilities and the introduction of DNHR will cause a regrouping of existing centers to manage the new networks which are forming.

4.1.2.2. No. 4 Electronic Switching Systems (ESS)

The facility which makes the continuing evolution of the interexchange network possible is the No. 4 Electronic Switching System, designed by Bell Laboratories Incorporated, and manufactured by Western Electric Company. In 1976, the first No. 4 ESS was placed in service in Chicago, Illinois. Since then, system hardware and software have been redesigned to take advantage of improvements in integrated circuits and high-level programming techniques. Systems in use today serve office sizes up to more than 60,000 terminations and may eventually switch up to 100,000 terminations, complete 500,000 busy hour calls, and switch over 40,000 erlangs.

Figure 4.13 shows the major subsystems of No. 4 ESS. Traffic enters and leaves through trunks at the left-hand side and is switched in the six-stage TSSSST network at the right-hand side. The four-stages of space switching are reconfigured 128 times every 125 μs. The time switches have 128 word buffers with sequential input and memory driven output (see Section 3.6). All signals are converted to T1 (see Section 3.5.2 and 4.2.3.2) format before connection to the digital interface unit. If required by the length of the connection, echo control is applied to the silent direction of transmission. A Special Services Unit provides mass announcements and other capabilities. The Stored Program Control evaluates the needs of incoming and outgoing trunks and the state of the system and

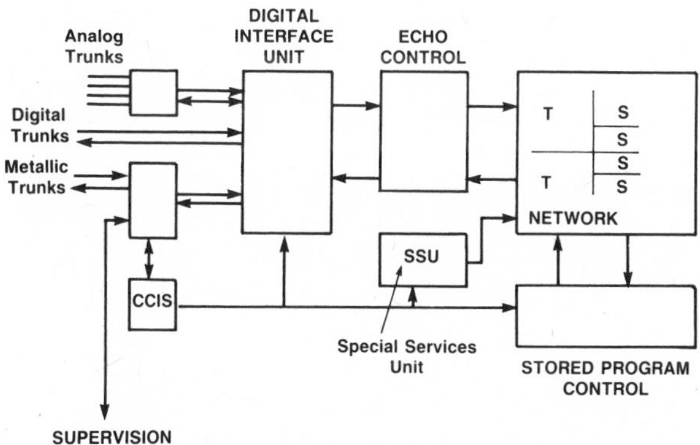

Figure 4.13. Major subsystems of No. 4 ESS.

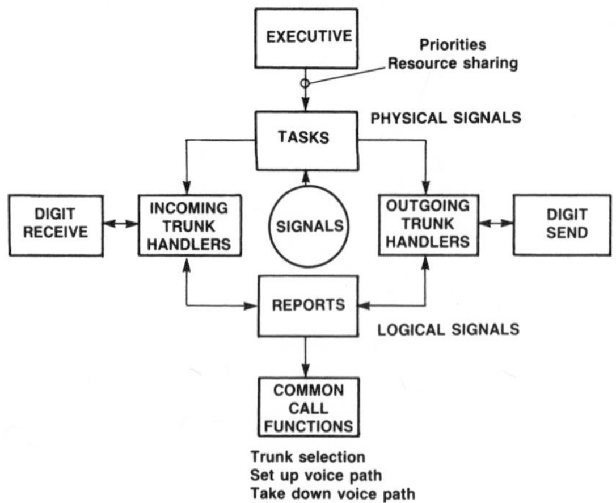

Figure 4.14. Representation of call processing in No. 4 ESS.

generates logical commands to facilitate the establishment of connections which create circuits between call*ing* and call*ed* parties.

The principle of call-processing is shown in Figure 4.14. Signals indicating requests for service and the status of circuits are received by the Tasks module and distributed to incoming and outgoing Trunk Handlers, which perform certain specific functions such as the receipt and logical encoding, and the logical decoding and sending, of digits, as well as determining the next step in call processing on the basis of the present state of the connection and the information contained in the signal received. This logical statement is transferred to the Reports module, which coordinates information on each incoming and outgoing pair of trunks. When appropriate, the module instructs the Common Call Functions module to perform a specific task directed to the setup or take-down of the voice path. The result is the passage of control data to the network or the signaling equipment.[(6)]

4.1.2.3. *High-Speed Switched Digital Service (HSSDS)*

In support of Picturephone Meeting Service (see Section 4.3.3), which employs 3 Mb/s digital video channels, HSSDS is available between major metropolitan areas of the United States. Implemented as 2 × T1 channels between selected switching centers, the service can be obtained on a reservation basis by those who have suitable T1 connections to these switches. A common network control center collects call setup and disconnect instructions and distributes them to the appropriate switching centers where calls are connected manually. In the future, an automated system will be introduced.[(7)]

4.2. Exchange Area Facilities

Exchange area facilities transport messages between persons in a local community of interest and pass long-distance messages to interexchange facilities. An estimate of the number of channel miles in use and the type of transmission systems employed is given in Figure 4.15.

4.2.1. Customer Loops

Telephones are connected to the local switch by means of pairs of wires which are called *customer loops*. Each loop is fashioned from wire pairs (twisted pairs) contained in a sequence of multipair cables which run from the serving switch to the user's premises. These cables are laid in ducts situated under roadways and sidewalks, buried directly in the soil, or suspended above the terrain between telephone poles. The sum of these facilities is referred to as *outside plant* (OSP). A continuing problem facing the outside plant engineer is to plan new feeder cables and provide relief to in-place cables so that the changing patterns of demand can be satisfied when they occur without an unnecessarily large investment in plant waiting to be used. Modern practice usually requires that feeder cables, which form the backbone of the network, are installed to meet forecasted demand over a planning period (usually seven years) and that distribution cables, which provide access from the feeder to the customer premises, are installed to meet ultimate demand.

In most areas, the local network is a mixture of methods and materials—the result of many years of change in telephone cables and outside plant construction techniques. Close to the switch, the cables are large and contain many fine copper wires. For very long loops the wire diameter may be increased in order to prevent the total resistance of the loop exceeding the value which just

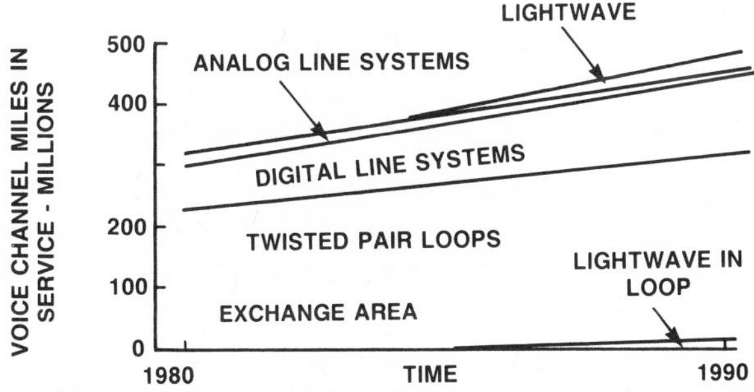

Figure 4.15. Estimate of total number of voice channel miles in service and the makeup of the types of facilities in use providing exchange area transmission in the 1980s.

allows sufficient current to flow for signaling. Known as *resistance* design, this technique is relatively costly both in terms of the initial investment in cable and the continuing expense of administering a local network with many gauges of wire. Since there is no point to being able to signal if the parties cannot hear each other, inductance is added to long loops to reduce the voice band attenuation of the connection. Known as *transmission* design, this technique inserts loading coils at fixed intervals to improve transmission of low- and middle-frequency voice energy at the expense of the higher frequencies (see Section 3.5.1). Thus, at 1000 Hz, the attenuation of 26 AWG loaded pairs is approximately 66% of the attenuation of 26 AWG unloaded pairs. For 19 AWG, the 1000 Hz attenuation of the loaded pair is approximately 33% of the unloaded pair. A loop must satisfy both resistance and transmission design parameters in order to provide satisfactory service to the user.

4.2.1.1. Unigauge Design

In order to reduce the cost of wire pairs in the loop plant, modern practice seeks to use as much fine wire (26 AWG) as possible. To do this, special arrangements are needed to serve the longer loops. One approach sometimes used is known as *unigauge* design. It employs 26 AWG pairs to serve all customers up to 30,000 ft aided by range extender equipment consisting of a higher voltage battery (to extend the resistance design limit) for long loops, and voice frequency amplifiers (to extend the transmission design limit). For the longest loops, 26 AWG loops are combined with a larger wire size and range extenders. (Range extenders are allocated to the loops which require them, when they require them, by the serving switch.) The details are shown in Figure 4.16. Up to loop lengths of 15,000 ft (which may serve as many as 80% of all subscribers in a representative area), 26 AWG nonloaded wire pairs can be used with a standard 48V battery. For loops between 15,000 and 24,000 ft, 26 AWG nonloaded wire pairs can be used with a 72-V range extending battery and voice frequency amplifiers. For loops over 24,000 ft, a combination of nonloaded and loaded connections is used with the 70-V battery and amplifiers. For loops between 24,000 and 30,000 ft, 26 AWG nonloaded wire pairs are used from 0 to 15,000 ft, and 26 AWG loaded wire pairs from 15,000 to 30,000 ft. For loops between 30,000 and 50,000 ft, 26 AWG nonloaded wire pairs are used from 0 to 15,000 ft, and a larger wire pair (22 AWG) is used with loading from 15,000 to 50,000 ft. The few loops beyond 50,000 ft must be specially engineered.

4.2.1.2. Concentrated Range Extension with Gain

Unigauge design is intended for general application to OSP connecting to end offices with electromechanical or electronic switches. For those offices containing No. 1/1A and No. 2/2B ESSs, range extension and gain circuits can be

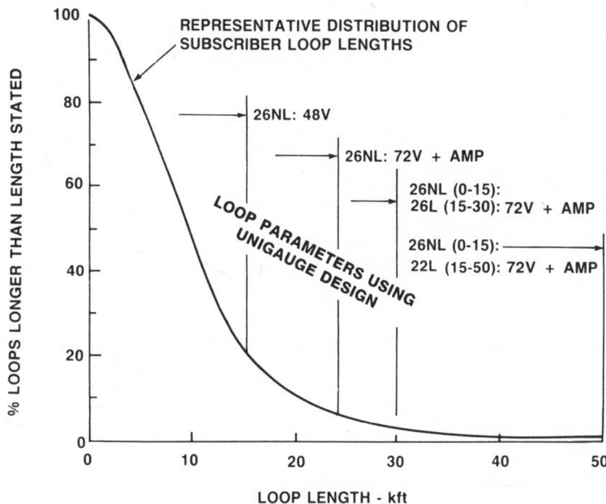

Figure 4.16. Distribution of subscriber loop lengths and loop parameters based on unigauge design using 26 AWG copper wire and a range extension battery. (26 = 26 AWG, 22 = AWG; NL = nonloaded, L = loaded; 48 V = standard battery voltage, 72 V = range extension battery voltage, AMP indicates the use of voice-frequency amplifiers).

embedded in the ESS network behind the first stage of switching concentration. Known as CREG (concentrated range extension with gain), the system allows signaling and transmission over loops up to 2800 Ω (approximately doubling the resistance of the loops which can be served). All loops longer than 15,000 ft must be loaded, and it is recommended that, when larger wire sizes are required for long loops, the loop be composed of at most two consecutive gauges (to reduce reflections). Locating the range extension equipment in the ESS network takes advantage of the inherent concentration to reduce the cost per line significantly.[8]

4.2.1.3. Serving Area Concept

For the most part, today's subscriber loops do not consist of straight through, permanently assigned connections between serving points and serving switch. In practice, from time to time, the physical connection will be changed as growth takes place, population shifts, and demands for service fluctuate. In some modern installations a concerted effort has been made to ensure that distribution cable is permanently assigned to each serving point and is isolated from the effect of changes in the feeder plant. Called the *serving area concept* (SAC), it is illustrated in Figure 4.17. Users are grouped into convenient areas, each of which contains a serving area interface (SAI) cross-connect frame. From one side, permanent connections are made through feeder cables to the serving switch. From the

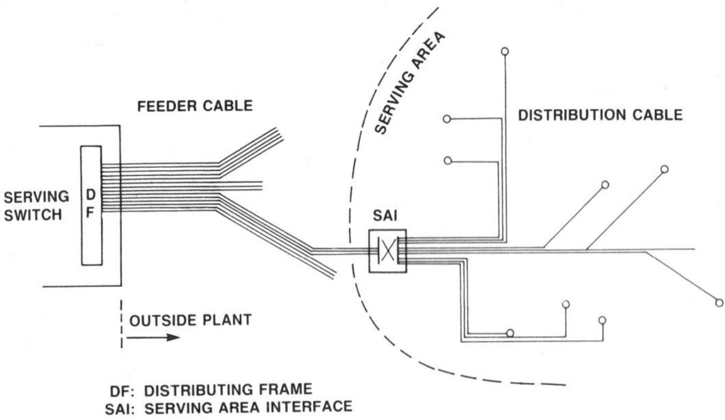

Figure 4.17. Representation of the Serving Area Concept (SAC).

other, permanent connections are made to the subscriber's premises. Installation or removal of service is effected at the SAI only. The degree to which this discipline can be followed depends on the availability of wire pairs, the current level of change activity, and the anticipated rate of growth. Thus, because of an immediate need to provide service for which permanent pairs are not available, some pairs may be reassigned from inactive serving points. Often, because the change may not be regarded as permanent, the old connection is *bridged* to the new one, leaving an active wire pair serving the new user attached to an inactive wire pair which served the former user. Such *bridged taps* (BTs) provide an additional load on the loop circuit which can cause high-frequency performance to deteriorate. In a typical situation, perhaps 20% of all subscriber loops may be without BTs, 40% may have one, 20% may have two, and 5% may have five or more. Their total effect depends upon the lengths of the active and inactive loops, the quality of the physical connections which have been made, and the proximity of the inactive loops to interfering sources which will produce noise.

4.2.1.4. Outside Plant Productivity

The customer loops and terminating telephones are the least used of all telephony facilities. On average, they are in use less than one hour each day. Further, while most pairs may be assigned to specific customer stations, there will always be a fraction of the plant which is waiting to be assigned to meet future customers requirements. A strategy to increase the throughput (or productivity) of the OSP is to shorten loop lengths by moving switching closer to the customer. This creates a requirement for smaller switches, and for short-distance multiplexing to carry the traffic to the end office. Other strategies are to introduce concentration or multiplexing at appropriate points where they can serve a community of customers. They replace a larger number of underutilized loops by a smaller number of loops which are in use more often. This effect is known as *pair gain*. In fact, all of these techniques are readily implemented in digital circuits and form the basis for a family approach to modern digital end offices in which a large base unit switch is supported by remote switching units, multiplexers, and concentrators. Further discussion of this equipment is reserved for Section 4.2.4. Their use can dramatically reduce the requirement for range extenders and coarse gauge wires.

4.2.1.5. Digital Signals

Fine wire, higher voltage signaling, voice frequency amplifiers, loading coils, and bridge taps have the primary objective of providing a local network which will support voice frequency communications at a reasonable cost. For all but low-speed applications, these techniques are harmful to the carriage of digital signals. In order to pass digital data, digital voice, and the higher bit rates required for two-wire connections to digital telephones, the loops must be *conditioned* (load coils and bridged taps must be removed), and intermediate repeaters may have to be installed. Conditioning existing outside plant for digital working can be an expensive and time-consuming task. In addition, two-way working can only be achieved with special arrangements. Over a single pair, either the use of the line must be time divided, so that terminal and switch use it alternately (time compression multiplex, TCM), or a hybrid with adaptive echo cancellation must be used. (Further discussion of these techniques is reserved for Sections 4.3.1.3 and 4.7.1, respectively.) Otherwise, two pairs are needed. If they are provided in the same small cable, near-end cross talk (NEXT) can be a problem, that is, interference between a high-level transmitted signal and an attenuated received signal both at the same end of the cable. Line codes such as WAL2 or CMI, described in Section 3.3.1.2, or partial response coding (Section 3.3.1.1), may be needed to overcome the high-frequency interference. If the systems can be synchronized so that there is never a time when a transmitter is transmitting, and a receiver is receiving at the same end of the cable, NEXT can be ignored. Far-end cross talk (FEXT), that is, interference between a receiver at one end of the cable, and transmitters at the other end, is less troublesome, but always present.

4.2.1.6. Digital Line Carrier and Optical Fibers

An alternative is provided through the use of digital line carrier (DLC) techniques and optical fibers which can be used to link to clusters of users who require either analog voice, digital data, or digital voice loops. Figure 4.18 illustrates the use of a DLC system which carries a mixture of analog and digital

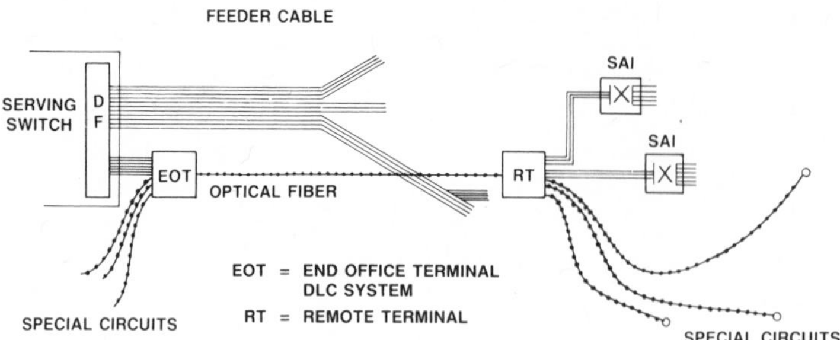

Figure 4.18. Application of Digital Line Carrier (DLC) and optical fiber to provide additional voice circuits, digital data circuits, and special high-speed circuits.

voice and data signals on an optical fiber. The signals are multiplexed in the end office terminal (EOT), carried on an optical fiber to the remote terminal (RT) where they are demultiplexed and distributed to SAI facilities, or carried on fibers directly to users who require special circuits. Customers with limited speed data requirements within a few thousand feet of the RT can be served through wire pairs. The widespread use of optical fibers in the customer loop, particularly from the RT to the customer's premises, cannot be said to be imminent. Substitution of this transmission medium for the existing wire pair loop would certainly improve channel quality and provide enough bandwidth for any conceivable service. However, the cost of fiber and associated support facilities is significantly higher than copper wire, and is likely to remain so—and there is always the question of what to do with all of the wire cables now in place.

4.2.2. Mobile Telephone

Providing mobile telephone service to a single subscriber (mobile) is a relatively simple radio application. Providing telephone service to many mobiles without the use of sophisticated techniques to reuse the frequency spectrum is impossible. In order to use a limited range of frequencies over and over again, a given area can be imagined to be divided into discrete regions, or cells, each of which is served by a specific band of frequencies. None of the cells surrounding a given cell employ the same frequencies, although cells separated by more than one cell from the given cell may use them. The level of isolation achieved between mobiles located in two cells which use the same frequencies depends on physical separation (cell size), power employed, and the actual frequencies used. Fortunately, in a typical urban environment, received power may vary with the third or fourth power of distance from the transmitter (not the second power as in free-space). This phenomenon is attributed to the blocking effect of buildings and local topography. It makes it possible to space transmitters closer together and provides sharper limits to the cells.

4.2.2.1. *FDMA Cellular System*

A frequency-division multiple access system can be used to provide high-quality mobile radio telephone service. Known as AMPS (Advanced Mobile Phone Service), the overall control of the system resides in a large central controller in each metropolitan service area. This Mobile Telephone Switching Office (MTSO) is an electronic switching system programmed to provide call-processing and system fault detection and diagnostics. The MTSO is connected directly by wireline to a set of base stations, called cell sites. Each cell site communicates to the mobiles through terrestrial radio stations. Small programmable controllers at each cell site perform call setup, call supervision, mobile locating, hand-off, and call terminating. A microprocessor within each mobile implements signaling, radio control, and customer alerting functions. This general arrangement is shown in Figure 4.19.

As developed by AT&T and approved by FCC, FM radio transmission from mobiles to cell sites uses frequencies between 825 and 845 MHz. From cell sites to mobiles, the frequencies used are between 870 and 890 MHz. Each radio channel consists of a pair of one-way channels separated by 45 MHz. The bandwidth is 30 kHz. Two types of channel are provided: setup and voice. The setup channel transmits and receives data used to set up all calls. The voice channel provides the talking path and also handles short bursts of data required for controlling and terminating the cells.

As the mobile moves from cell to cell, the MTSO arranges for the talking path channels to be switched to those available in the cell just entered. This is

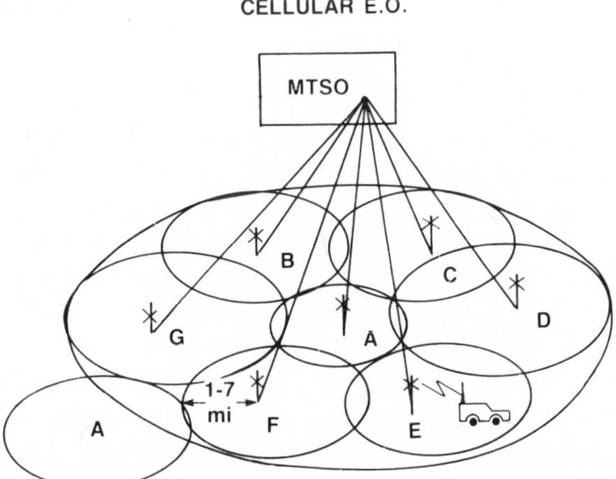

Figure 4.19. Representation of cellular radio mobile telephone system showing radio connection from mobile to cell site controller and wireline connection to end office. Frequency band A is reused beyond ring of cells using bands B—G.

done when the cell site reports that an in-progress call has dropped below an acceptable signal level. The MTSO instructs all cell sites surrounding the reporting cell site to monitor the setup channel. From the reports received, the MTSO determines which cell site is closest to the mobile and instructs both cell sites and the mobile to prepare for hand-off. At a given instant, the wireline connection to the first cell site is broken and a connection made to the second cell site, where the stationary end of the talking path is established through a transmitter/receiver pair operating on different frequencies from those used in the first cell. At the same time the mobile retunes to the new frequencies to complete the new channel. The entire process takes place in milliseconds so that the mobile user perceives no interruption in the call.

At present, wireline and nonwireline carriers share the frequency allocation so that there can be two mobile telephone services in the same area. Each can serve up to 333 mobiles per cell. In a metropolitan area divided into 16 cells, it is estimated that satisfactory grade of service can be provided to some 100,000 mobiles. For cells in which the channel capacity is exceeded, smaller cells can be created, replacing one large cell (7–8 miles radius) by seven smaller cells (1–2 miles radius), for instance. This requires the use of lower-power transmitters at the new cell sites.

4.2.2.2. *Single-Sideband AM*

Narrowband FM is an established technique which is well understood and for which techniques and components are readily available. It is not, of course, the most efficient analog modulation which could be employed. That distinction belongs to single-sideband AM (SSB-AM) which requires no more than 4 kHz per voice channel. In fact, since voice has been found to contain mostly low frequencies or mostly high frequencies, but not both together, it is possible to fold the spectrum to create a composite signal of around 2 kHz bandwith. Using this frequency companding technique, SSB-AM requires approximately 2.5 kHz per channel, making it more than ten times as efficient as narrowband FM. However, frequency companding can degrade voice quality. Further, FM is relatively insensitive to interference: the same cannot be said for SSB-AM. For these reasons, immediate application of this technique to mobile telephone service is unlikely.

4.2.2.3. *Spread-Spectrum System*

Instead of using FDMA techniques, cellular mobile radio systems can employ CDMA (code division, multiple access) techniques. In this spread-spectrum arrangement, more users can be accommodated and there is the possibility of operating in bands assigned to other services.[(9)] A direct sequence spread-spectrum system transmits and receives a code whose sequence is known to both the transmitter and receiver. The code occupies a much wider bandwidth than the information carried. There may be 100 to 1000 (or more) code bits (or chips) for each information bit. The more chips, the more spread the spectrum, and the less interference between mobiles and with other spectrum users. For a system employing a 100 to 1 code rate (i.e., a processing gain of 20 dB) and 9.6 kb/s digital voice, the bit rate will be approximately 1 Mb/s and the RF bandwidth will be around 0.5 to 1 MHz depending on the modulation technique employed. (Spread-spectrum modulation is discussed in Section 3.3.5.)

Spread-spectrum techniques have been estimated to improve on narrowband FM by factors of 10 to 100 times (on the basis of users per MHz). They can be used in the presence of strong interference and have been shown to be able to operate over frequency bands containing narrowband transmissions without harm to any of the signals. In portions of the spectrum which are underutilized by narrowband systems, the responsible operation of a spread-spectrum overlay will have no harmful effect. For parts of the spectrum which contain many active narrowband emissions, the effect of a spread-spectrum overlay has yet to be determined. Spread-spectrum technology is evolving. Realistically, it is a candidate for a future generation of mobile telephone systems, not the current system.(10)

4.2.2.4. Personal Radio

The cellular mobile radio (CMR) systems described are too expensive (several thousand dollars per user) for a personal radio service (PRS). By sacrificing wide-area operation, limiting each person to operating within a maximum distance (five miles) of a single assigned base unit, and connecting this unit to the telephone network, a limited personal communication capability could be supplied at a projected price of a few hundred dollars.(11) No system is presently in place to do this.

4.2.2.5. Satellite Land Mobile Radio

Rural needs for mobile radio service can be supplied from a geostationary satellite fitted with an extremely large antenna which would enable it to communicate with small, simple, inexpensive terrestrial equipment. A call originating from a home telephone would be routed to an earth station through the public switched network. It would be transmitted to the satellite at C or Ku bands and retransmitted to earth to the mobile receiver in one of two bands between 800 and 900 MHz. The second band would be used to carry the reply to the satellite and thence to the caller through the earth station and telephone network. The system could meet the needs of law-enforcement, emergency medical services, interstate transportation, and other uses located beyond the range of cellular mobile systems based in urban and suburban areas.(12) It is unlikely that systems of this sort will be available in the near future.(13)

4.2.2.6. Application to Rural Telecommunication

Sparsely populated rural areas are difficult and expensive to serve. They are characterized by scattered, low-density population; scarcity of infrastructure (such as electric power and transportation); scarcity of human resources for operation and maintenance of telecommunication equipment; hazardous topology (perhaps); severe climate (perhaps); and a low level of economic prosperity (usually). All of these factors conflict with the provision of conventional wire facilities in these areas. They may be served more easily through a chain of radio cells which provide wide-area coverage and employ high-performance digital equipment small enough to be solar powered. Signals are passed from a central switch to a cell site and then from cell site to cell site in digital multiplexed streams. At each site, local calls are dropped and radiated within the cell on different frequencies. The return channels are received and inserted into the return multiplexed stream. If necessary, some cell-to-cell connections can be made through satellite facilities.[(14)] Connections within one cell can be made by the cell site equipment if switching capability is provided there.

4.2.3. Intraexchange Transmission

Between end office switches in an exchange area, most traffic is transported on wire cables. Both analog and digital techniques are used to multiplex 12, 24, or 48 voice channels over distances from a few miles to tens of miles. The nonloaded cable pairs are made from relatively large diameter copper wires (19 or 22 AWG) and are connected so as to provide separate circuits for the two directions of transmission. Some traffic is carried by microwave radio, and an increasing amount is carried on optical fibers.

4.2.3.1. FDM Carrier Systems

In the United States, frequency-division multiplex is used in *N*-carrier systems which carry twelve or twenty-four 4-kHz voice channels on wire pairs between local exchange facilities. To make them cost effective, even over a few miles, simple terminals and straightforward modulation techniques are employed at the expense of bandwidth and power. Thus, *N*2 equipment, which was introduced in 1962, uses double-sideband, transmitted carrier, amplitude modulation to place 12 one-way channels between 36 and 140 kHz (low group) and 12 one-way channels between 164 and 268 kHz (high group). In operation, the go pairs are assigned to one group and the return pairs are assigned to the other. At each repeater the frequency assignments are interchanged so that all repeater outputs (for both go and return pairs) are in one band, and all repeater inputs are in the other. This frequency separation minimizes cross talk and provides equalization. For longer systems, terminal cost can be increased in order to gain more channels. *N*3 equipment, which was introduced in 1964, places two groups of 12 single-sideband channels in both high- and low-frequency bands to provide a 24-channel system.

4.2.3.2. *TDM Systems*

In the United States, time-division multiplex is used to transmit 24 and 48 voice channels on wire pairs between local exchange facilities. Even though no *carrier* is used to derive the signal, they are designated T1 and T1C carrier systems, respectively. (In many ways, the fundamental frequency at the bit rate is the carrier.) In a T1 system, analog-to-digital conversion and multiplexing is performed by equipment which encodes 24, 4 kHz voice channels into a 1.544 Mb/s stream for transmission over a wire pair. One sampling period (125 μs) contains 24 8-bit words (192 bits) plus a *framing* bit (193rd bit) which allows the receiving section of the channel bank to identify the sample sequence. Thus the bit rate is 1.544 Mb/s, not 1.536 Mb/s (24 × 8 bits × 8000). However, not all of the bits in each 8-bit word are used for voice information. Either the last bit of each word in all frames is used for signaling, or the last bit in each word of every sixth frame is used for signaling. For data applications, the presence of either technique limits the maximum information rate per channel to 56 kb/s.

T1 systems were first introduced in 1962. T1C systems which multiplex 48 voice frequency channels into a 3.152 Mb/s stream were introduced in 1975. Today they account for a large fraction of the growth in the metropolitan trunk plant. In all, T-carrier provides around half of all metropolitan interoffice connections.[(15)]

4.2.3.3. *Microwave Radio*

4.2.3.3.1. Analog Systems. For the shorter, intraexchange area distances, analog microwave radio has significant competition. Nevertheless, microwave radios which employ frequency modulation (FM) and which carry frequency-division multiplexed (FDM), single sideband, suppressed carrier voice channels, are used extensively between local exchanges where distances or ease of construction warrant.

4.2.3.3.2. Digital Systems. As discussed in Section 4.1.1.1.2., the cost per circuit for digital radio is less than analog radio, for distances up to several hundred miles. As a result, digital radio is the facility of choice for route relief and new construction in the exchange area when radio is indicated.

4.2.3.4. *Optical Fiber Systems*

As a replacement for individual 12-, 24-, and 48-channel wire pair transmission systems, optical fiber systems are uneconomical. However, for the 150 working systems contained in a 900-pair cable (150 × T1 = 3600 two-way voice circuits: 150 × T1C = 7200), a few fibers (12 or 24 fibers equipped at FT3, for instance) can make an ideal replacement. As the technology improves, all intraexchange, inter-end-office, optical fiber system connections will be completed without using repeaters thereby improving network reliability and reducing operating costs. Further, in areas in which underground ducts are used, the substitution of fiber cable for copper cable will free significant space for future expansion. Where they can be used, optical fiber systems are rapidly becoming the medium of choice for route expansion and for new installations. The combination of optical fiber, digital radio, and digital switching, virtually assures all-digital transmission between end offices as rapidly as existing equipment can be retired.

4.2.4. Exchange Area Switching

An exchange area consists of a set of end offices (EOs) which interconnect individual customers and which are themselves interconnected to provide calling to neighboring EOs.

4.2.4.1. Network Architecture

Each EO connects with a tandem switch which acts as the access point to interexchange facilities (Access Tandem, AT). These arrangements are illustrated in Figure 4.20. Switch manufacturers are making maximum use of modularity and commonality in both hardware and software to provide end-office switches which can accommodate a wide range of applications (from small rural communities to large metropolitan areas), and to facilitate graceful evolution to future performance which can satisfy the needs of an information-based society. High on the list of requirements is a design which includes a modular family of units that places pair gain as close to the customer as possible through the use of concentration, multiplexing, or remote switching (for the reasons given in Section 4.2.1.4).

4.2.4.2. GTD-5EAX Switching System

The GTD-5EAX (Electronic Automatic Exchange) is an example of such a family. Designed and manufactured by GTE Communication Systems Corporation, it provides exchange area switching spanning office sizes between 500 and 150,000 lines. It consists of a Base Unit (BU) which serves three types of remote units: Remote Switch Unit (RSU), Remote Line Unit (RLU), and Multiplexer Unit (MXU). The base unit is implemented in two sizes. The Large Base Unit (LBU) has a size range of 2000 to 150,000 lines served by the base itself or subtending remote units, a traffic capacity of over 30,000 erlangs, the

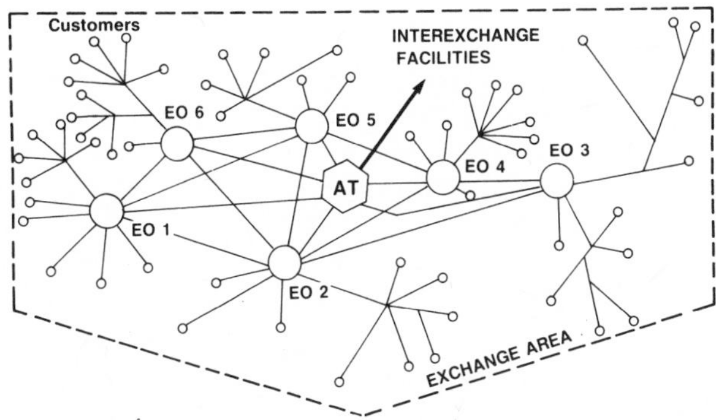

Figure 4.20. Exchange area configuration showing six end offices (EOs) connected to customers, interconnected to provide calling to neighboring EOs, and connected to an access tandem (AT) switch which connects to interexchange facilities.

capability to service 360,000 call attempts per hour, and memory to store 300,000 directory numbers and associated information. The LBU connects to all three remote units by digital T1 facilities. The Small Base Unit (SBU) shares a common architecture with the LBU. It has a size range of 500 to 20,000 lines, a traffic capacity of over 4,000 erlangs, and the capability to service 65,000 call attempts per hour and support 40,000 directory numbers. The RSU is a small switch designed to serve outlying communities and as a replacement for community dial offices (CDOs). It serves up to 3000 lines with a capacity of over 500 erlangs. Controlled by either size base unit, it provides local-to-local switching and can host both RLUs and MXUs. The RLU is a concentrating pair gain unit serving up to 768 lines with over 100 erlangs capacity. The MXU is a concentrating, or nonconcentrating, pair gain unit serving up to 96 lines.[16]

4.2.4.2.1. Hardware Architecture. The hardware architecture of the GTD-5EAX is illustrated in Figure 4.21. It consists of peripheral, network, and control equipment in the base unit, and remote units with individual microprocessors. The BU consists of a range of Facility Interface Units (FIUs) which present standard interfaces to the switching network and control equipment. The network is a time-division multiplexed, full availability, essentially nonblocking, time–space–time network, switching 12 parallel bits in each of the 64-kb/s channels. For reliability, it is fully duplicated and both synchronized copies switch identical information. The RLU consists of a Remote Switch and Control Unit (RCU) module interconnecting a maximum of four analog line and/or digital trunk FIUs (terminating either lines or digital links from subtending RLUs and/or MXUs) to a maximum of four digital trunks connecting to the base unit. The RCU contains the same network elements (originating and terminating time switches, digital tone sources, and digital pads) as the RSU.

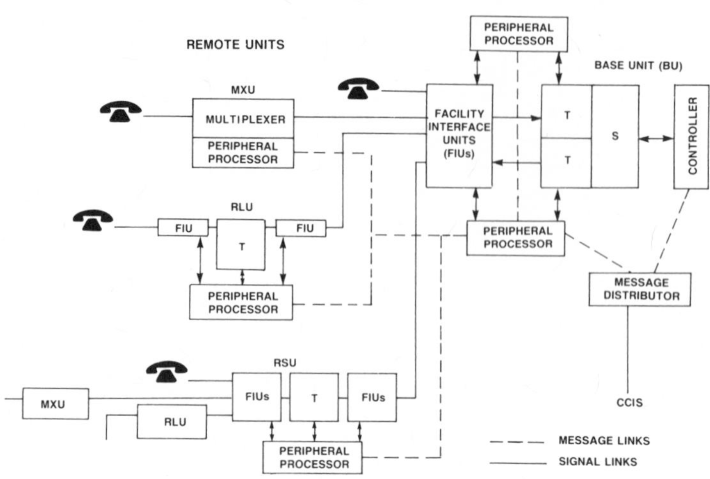

Figure 4.21. Block diagram of GTD-5EAX system. It consists of peripheral, network, and control equipment in the base unit, and remote units with individual microprocessors.

The control architecture is modular, distributed, and multiprocessing. It consists of a central control and three levels of peripheral processors. A peripheral processor unit is allocated to specific hardware and its program performs functions peculiar to the hardware it controls. The central control processor units perform the logical analysis and sequencing of calls. Each consists of an administrative processor unit used for the administrative and maintenance functions, and up to seven identical load-sharing telephony processor units used for call-processing functions. Central control processor units and base unit peripheral processor units communicate by way of a duplicated message distribution circuit. It provides a rotational priority and conflict resolver circuit to arbitrate contention between the processor units. Remote processor units communicate to a peripheral processor unit or other remote processor units over 64-kb/s data links.

4.2.4.2.2. Software Architecture. The overwhelming majority of the software modules are written in an extended and enhanced version of PASCAL. Real-time modules requiring very efficient execution are written in assembly code. The software is organized into five hierarchical levels: program, class, subprogram, module, and segment. The program is the highest level and includes all on-line software with each succeeding level being a functional breakdown of the level above it. The program is broken down into 12 classes with each class being a major functional grouping. The subprogram is a functional breakdown of the class. Program, class, and subprogram are used for control and documentation. There is no procedural code at these levels although the data base is organized around these partitions. Data common to a subprogram are placed at a level in the hierarchy so that only modules in that subprogram can be authorized to access the data. The module is the compilable, configuration-controlled entity in the system. Modules are individually compiled and later linked into a complete load. The segment is the lowest level in the organization and most do not exceed about 50 source statements.

4.2.4.3. Other Digital Switches

The GTD-5EAX is one of several digital end-office switches which have been introduced in the last few years. All are digital, time-division multiplexed switches which operate in conjunction with 24-channel T1 trunks or 32-channel CCITT trunks. They all employ TST networks. To various degrees they support remote options and have significant modularity in both hardware and software. Within each product line, common hardware and software are employed to the greatest degree possible. Most software is written in a high-level language and is transportable and extendable. All use IC semiconductor memory (MOS or CMOS). For the operating company, a digital switch family offers the opportunity to extend switching and/or concentration to small communities of subscribers, as well as the opportunity to consolidate small existing end offices (EOs) into a large base unit. This flexibility can result in significant savings from reduced building requirements (fewer EOs) and reduction or elimination of growth in feeder plant. Further, the all solid-state implementation requires far less maintenance, and the all electronic design makes it possible to use auxiliary computer-based systems to supervise the functioning of several end offices, automating diagnostic testing and trouble reporting, and allowing automatic recovery from most failures. Today's digital switches occupy less space (per line) than analog ESSs or electromechanical switches and require less power. Through architectural and functional evolution, tomorrow's digital switches will occupy even less space, consume even less power . . . and cost even less (on a per line basis). As integrated circuit chips increase in complexity, entire systems/subsystems will be placed on a chip and even the largest base unit will become a few chassis in a rack of equipment attached to optical fibers which connect to remote units located close to the customers.

4.2.4.4. Working in a Mixed Environment

In today's mixed analog and digital environment, interface equipment is required to bridge between the digital base unit or remote units and the wire pair line connected to the customer's telephone. Several functions must be performed: they are often referred to as BORSCHT, which is an acronym for battery feed, overvoltage protection, ringing, supervision, coding and filtering, hybrid, and testing. *Battery feed* supplies the current for signaling and powering the customer's instrument. *Overvoltage protection* provides protection for the electronic interface circuits against the high voltage surges induced by lightning or accidental contact with other electrical sources. *Ringing* applies a relatively high, modulated voltage to ring the customer's instrument. *Supervision* detects the existence of a low impedance at the customer's instrument following pickup of the handset (off-hook condition). *Coding and filtering* provides for analog-to-digital (A/D) and digital-to-analog (D/A) conversion of the voice signal, including filtering (F) for antialiasing and reconstruction. *Hybrid* provides the physical conversion between the two-wire, two-way circuit of the customer loop and the two, one-way connections to the switching equipment. *Testing* provides for the automatic testing of the customer loop from the end office. A typical BORSCHT arrangement is shown in Figure 4.22.

Interface equipment will also be needed to connect the digital switch to trunks. Although the particular tasks may be somewhat different, some of the basic functions performed on the customer side must also be performed on the trunk side. Thus battery feed, overvoltage protection, and line testing are performed by the trunk interface equipment. In addition, codecs and filters are required for analog trunks and a hybrid is necessary for two-wire analog trunks. Most digital trunks can be connected without signal processing since they employ the digital signals commonly used in digital switches.

Most of the BORSCHT functions are being implemented in integrated circuits. Thus a subscriber-line interface circuit (SLIC) which consists of supervision; coding, decoding, and filtering; and hybrid conversion, is available on a single IC chip. The other functions of battery feeding, ringing, and testing involve higher voltages and currents, and overvoltage protection requires the ability to withstand very high voltages (typically 1500 V). Significant progress is being made in high-voltage silicon ICs, and integration of the entire set of BORSCHT functions is only a matter of time. When this occurs, it will produce a significant reduction in the cost of a digital switch.

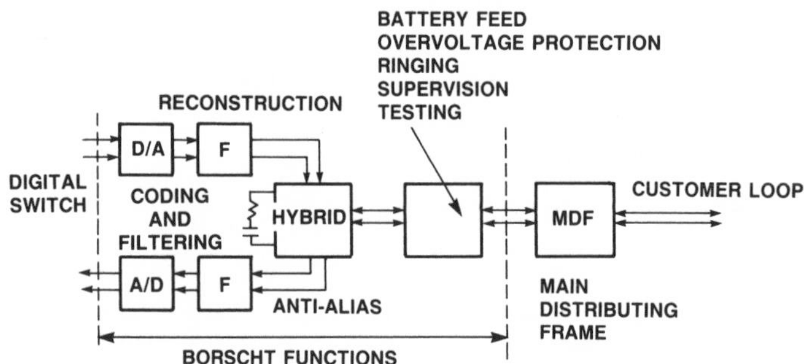

Figure 4.22. Representation of the location of BORSCHT circuits between a digital switch and a customer loop connected to an analog telephone.

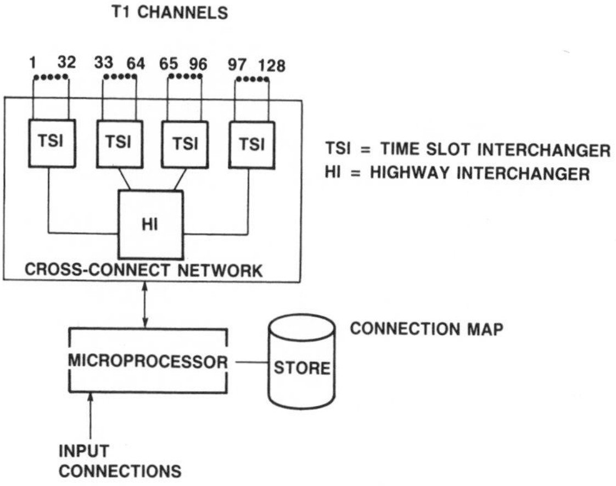

Figure 4.23. Principle of the digital access and cross-connect system (DACS).

4.2.4.5. Digital Access and Cross-Connect System

As digital end offices are deployed, RSUs, MXUs, and RLUs are introduced in appropriate locations in the exchange area, and customers install digital PBXs and advanced equipment, T1 carrier facilities will proliferate. A system called DACS (digital access and cross-connect system) is used to consolidate these facilities and to segregate channels by usage and destination. In this way, variations in the traffic patterns encountered in exchange area digital trunks can be accommodated without (necessarily) installing new trunks. DACS provides the ability to directly cross-connect digital voice channels residing in T1 digital streams in a universe of up to 127 streams. A 128th stream is reserved for test access. In all, 3048 individual 64-kb/s channels, or 1524 two-way circuits terminate in a DACS frame.

A block diagram of DACS is shown in Figure 4.23. DACS consists of four time-slot interchange (TSI) units connected by a digital space switch (HI). Connections are set up and taken down by a microprocessor under the control of inputs from terminals operated by maintenance staff. A map of the connections is stored permanently so that they can be reestablished in the event of a failure.

4.3. Customer Premise Equipment

Customer premise equipment is used to process and distribute messages within households, office buildings, commercial complexes, and other institutions. It consists mainly of telephones and PBXs (and associated equipment).

4.3.1. Telephone

Over a hundred years ago, the principle of the telephone instrument was discovered and the familiar desk set with dial, carbon microphone, and electromagnetic earphone has been in existence for more than 50 years. Steadily, improvements have been made in the materials employed and the manufacturing processes used to produce a reliability which ensures operation under almost all conceivable circumstances. The basic voice telephone may be the most thoroughly engineered device the world will ever know. Despite this, it has outlived its usefulness as a modern telecommunication terminal, and its support imposes requirements on the remainder of the plant which are incompatible with the service and equipment improvements made possible by digital technology.

4.3.1.1. Dial Telephone

The elements of a dial telephone are illustrated in Figure 4.24. Ringing is accomplished by applying a voltage at the end office to the pair of wires serving the telephone. This activates the bell. By lifting the receiver, the person answering the call closes the hook switch, which stops the ringing and places the speak and hear circuits on the line. This change of impedance indicates to the end office that the call has been answered and that the connection should be made. The voice of the answering party activates a carbon microphone, which modulates the current passing through it, sending a signal along the line to the caller. Because the microphone is connected at the center of the balance (hybrid) transformer, most of the answering voice signal is balanced out, and does not reach the earphone. The small fraction of the signal which does, produces a sound known as *sidetone*. At the proper level it provides a balance to the sound entering the user's other ear over the air path between ear and mouth. The result is a

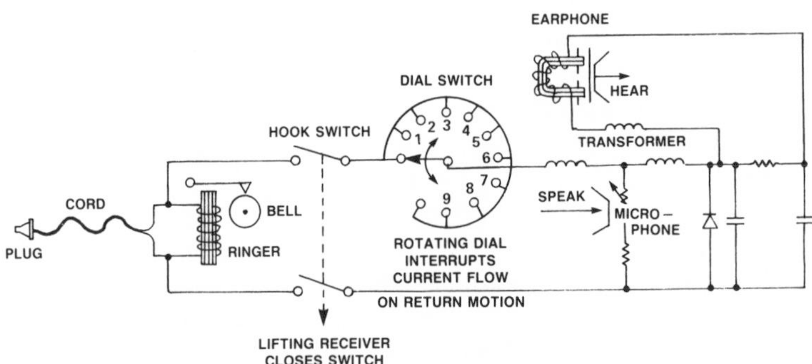

Figure 4.24. Representation of the elements of a dial telephone.

natural *feel* to the conversation. The voice signal from the caller is connected directly to the answerer's earphone through the transformer windings.

To initiate the call, the caller lifted the receiver, thereby closing the hook switch and causing current to flow from the central office. This flow of current was detected at the office and dial tone was sent indicating the office was ready to receive the called number. The caller dialed the digits. This action interrupted the current as the dial switch opened and closed while the dial returned to its resting position. When the number was received at the end office, the loop serving that number was tested and, if not in use, ringing voltage was applied.

As discussed in Section 4.2.1, because the action of the telephone depends on the flow of current from the central office, there is a limit to the distance from the office at which it may be operated. This can be calculated from the maximum resistance of the *loop* which will just allow the minimum operating current to flow under the influence of the voltage available in the office. Since it is not practical to adjust all loop lengths to the same resistance, telephone sets must operate over a wide range of currents. Further, because the wire connections are susceptible to interference from man-made sources, as well as lightning strikes and electrical storms, the telephone set must be able to withstand a wide range of voltages. A measure of protection is usually provided by attaching lightning arresters and fuses at the point at which the telephone connection enters the subscriber's premises. This uncertain environment has mitigated against the application of solid-state electronics. Despite this, telephones employing electronic circuits and digital techniques are rapidly replacing that electromechanical masterpiece, the dial telephone.

4.3.1.2. Electronic Telephone

Today's electronic units employ solid-state circuits to perform all of the usual telephone functions, as well as new ones, and to be compatible with existing telephone plant. Thus, they are two-wire, generate DTMF or simulated dial-pulse signaling, and incorporate electronic voice transducers. In many models, these transducers also serve as the ringer: the well-known bell sound has been replaced by an electronic warbling.

The substitution of electronic circuits for electromechanical parts has reduced the size and volume, making new shapes possible. Many electronic telephones consist of a single unit of about the same size as the standard handset (this being required to span the distance between the user's ear and mouth). The use of integrated circuits makes it possible to incorporate additional features. Some perform the traditional functions better: for instance, circuits can be included to make acoustical performance independent of line length. Others perform additional functions: for instance, many electronic units incorporate a recall feature which, upon request, repeats the last number called—a convenience if the number was busy or the caller forgot to say something. In addition, electronic units have made answering possible from virtually anywhere in the household or office through the use of low-level radio or infrared links and a portable telephone unit powered by a battery.

4.3.1.3. Digital Telephone

To this point, the units discussed employ analog speech. As digital techniques penetrate all levels of telecommunication facilities, interest is growing in units which employ digital speech. The most difficult function to implement is two-way (duplex) digital transmission over a pair of wires. In the analog case, this is relatively easy to achieve through the use of the hybrid transformer. But digital voice consists of pulses at a 64-kb/s rate (or something less if more complex encoding is employed). A line code which achieves 1-b/s/Hz coding efficiency requires that transformer balance be achieved over 64 kHz bandwidth—a much more difficult task than the 4 kHz bandwidth required of circuits which handle analog voice. It is further complicated by the frequency dependence of the line characteristics and the necessity to achieve a good impedance match between the balance network and the line for the wide variety of loop lengths and line conditions encountered in practice. The full power of a signal-processing chip can be put to work to provide a combination of adaptive hybrid and echo canceller. At the moment, this is an expensive solution. So too is a solution which employs two, two-wire connections, one for voice inbound to the telephone and the other for voice outbound from the telephone (see Section 4.2.1.5). Another alternative employs FDM techniques to support baseband transmission in one direction and carrier operation in the other. However, this marriage of analog and digital techniques cannot take advantage of the whole arsenal of digital technology.

By far the most popular arrangement to achieve duplex operation over a pair of wires employs a version of TDM which has come to be identified as time compression multiplex (TCM). Each direction of transmission is assigned a series of time slots which are interleaved. The principle is illustrated in Figure 4.25. Digital bursts of information from the far end alternate with bursts of information from the telephone. Because the line is shared between the two directions of transmission, the data rate must be at least twice the one-way rate. In fact, it is even higher because dead time must be allowed between the bursts so that they can be distinguished clearly and an allowance must be made for propagation delays along the wire pair (approximately 5 μs/km or 8 μs/mile). The electronic circuitry embodies buffers in which words enter at one rate, are combined with other data, and leave at another rate in order to balance the requirement for end-to-end operation at 64 kb/s. The word length will be greater than the basic 8-bit PCM voice sample since a header block is needed to identify origin and may contain signaling information.

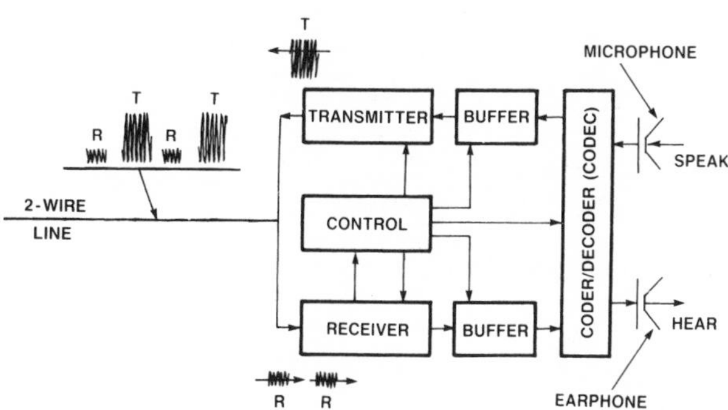

Figure 4.25. Illustration of the operation of a digital telephone which employs time compression multiplex (TCM) to achieve two-way transmission over a single pair of wires.

In analog telephony, signals sum easily so that the provision of extension telephones requires only that they be connected in parallel with the main station. In fact, there may be no operating distinction between the various instruments on the line. The number which can be used simultaneously is limited by the ability of the loop circuit to provide sufficient current to activate the units and to provide sufficient audio power to each instrument so that the caller can be heard. In digital telephony the operating conditions are quite different. Adding signals which represent binary words does not yield the sum of the binary words: they must be combined according to the rules of binary arithmetic. Connecting digital telephones in parallel, then, will not serve to implement extensions. In fact, each telephone on the circuit must be provided its own time slot and the collective sounds of all the talkers must be constructed logically at some location and fed to the users. Thus, the number of digital telephones which can be operated in parallel is limited by the data rates. Setting the rate specifies the maximum number of extension telephones which may be employed simultaneously, and defines a regime of line lengths over which the equipment will operate. For every user, received volume will be the same since the signal level is contained in the coding, not in the amplitude of the signal (as in the analog case).

The principle of three digital telephones operating in parallel is shown in Figure 4.26.[17] The system operates at a line rate of 512 kb/s. Besides 8-bit voice information, the words contain extra bits for ringing, supervision, and testing. This allows the user to talk and signal simultaneously. In one period, chosen as 125 μs to be compatible with a 64-kb/s voice sampling rate, the word from the calling party is broadcast to all telephones. Each telephone replies in a separate time slot. With the guard times listed in the diagram, the serving loop cannot exceed 2.2 km (1.4 miles), and the individual telephones cannot be spaced more than 400 m (one-quarter mile) apart. While these restrictions may not be arduous, they illustrate a limitation of TCM which does not affect implementation by adaptive hybrids and echo cancellers.

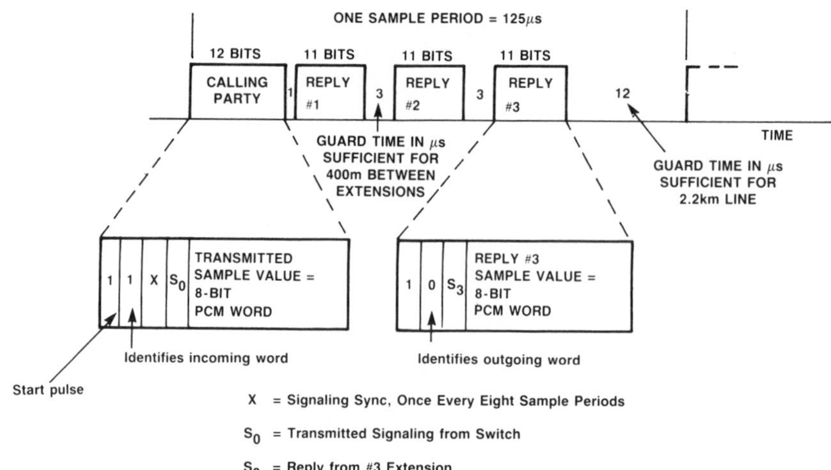

Figure 4.26. The message format of a digital telephone with extensions.

Figure 4.27 shows a block diagram of a digital telephone. AMI, or other bipolar coding is used on the line. The synchronizing bit in the inbound word maintains the state of the clock in synchronism with the incoming digital stream on the line ensuring that the logical operations occur at the proper time. The signaling bits from the successive words are accumulated, decoded, and used to

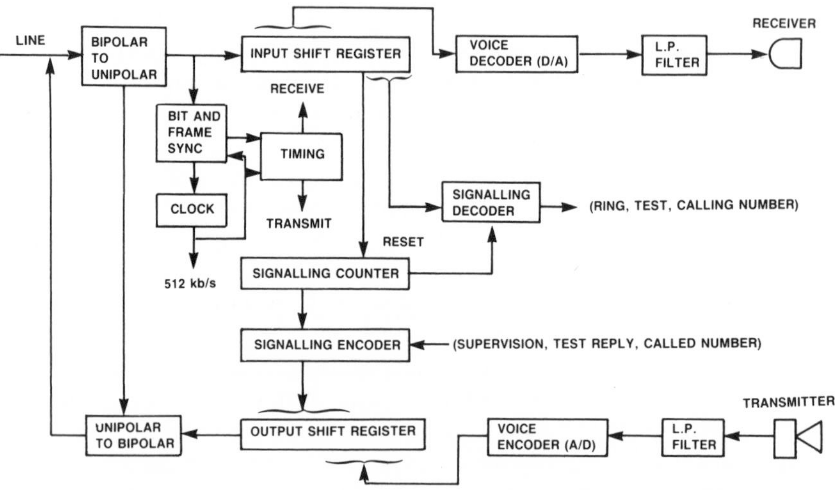

Figure 4.27. Block diagram of digital telephone.

initiate the required activity. The incoming voice word is converted to analog and applied to the receiver. Sidetone can be produced by supplying pulse amplitude samples of the transmitted voice at a reduced level to the low-pass (LP) filter ahead of the receiver. The system is synchronous in the sense that each telephone extracts and conforms to the bit rate of the inbound word. However, the arrival time of the words from the other extension units will vary with the distance separating them, making a combination of synchronous and asynchronous techniques necessary if conversation combining is to be achieved in the telephone unit. Operation will be simplified if this function is performed at the local switch. Here the extension replies can be processed and the sum, including any signal from the calling party, returned to each telephone in the normal inbound time slot.

4.3.1.4. Cordless Telephone

Since its invention, the telephone has consisted of a base unit anchored to the wall or placed on a desk, and a tethered listen-only, or listen and talk, unit. With the development of solid-state electronics, the listening, talking, and signaling unit can be separated from the base unit and operated up to several hundred feet from it. The channel between them can be supported by radio frequency or infrared (IR) radiation. Most cordless telephones use FM modulation and operate with radio carrier frequencies around 49 MHz. The use of IR is discussed in Section 4.6.3.

4.3.2. Digital Private Branch Exchanges

Office buildings, factories, hotels, and hospitals have a significant amount of voice traffic which stays within the institution. It is often switched by a private branch exchange (PBX) which connects internal users together and provides shared access to the serving end office. Competition for the sale of this equipment has forced manufacturers to design products with a large number of special features. In today's digital PBXs the full power of computer control is put to work to provide extra customer services as a way of differentiating competing systems. Thus, a hotel/motel version will provide mechanized wake-up calling, report use of pay-TV channels, monitor room status, supervise smoke detectors, etc. as well as the more conventional functions of calling card verification, automatic room number identification, message detail recording, call billing, etc. Features which are likely to be found in versions intended for business use include call forwarding, camp-on, call-waiting, conferencing, restricted calling, most economical routing, etc. These features may also be provided by equipment associated with the end office. Called Centrex service, it is supplied by the local operating company.

The first digital PBXs were introduced in the mid-1970s. At that time, the equipment consisted of a digital time switching network controlled by a minicomputer or microprocessor. While switching and control were fully digital, external connections were fully analog (with the exception of the possible use of T1 systems to connect to the end office). Today's equipment has benefitted from a generation or two of development spurred by a demand for office automation. Electronic message services, voice mail, protocol conversion, data switching, local area network interfaces, the integration of voice and data, support of smart work stations, etc. are some examples of the functions which modern equipment may incorporate. Further discussion is reserved for Section 5.1.3.

4.3.3. Picturephone Meeting Service (PMS)

The human aspects of videoconferencing have been discussed in Section 2.3.3.4. PMS is a video teleconferencing service offered by AT&T which provides two-way interactive audio and video between any two compatible conference rooms that are connected to the HSSDS network (see Section 4.1.2.3.). The NTSC color video signal and analog audio signals are digitized, compressed, encrypted, and combined with control signals to produce two T1 digital streams which are transported by HSSDS to the other conference location.[18]

B. TELEVISION AND RADIO

In this section, we discuss changes in television and radio, particularly those areas in which new technologies are being employed to improve performance. Not all are digital—but they provide increased dimensions to the residential telecommunications experience. The facilities described provide one-to-many (broadcast) video and audio connections as well as point-to-multipoint connections.

4.4. Television

The historical development of television services has led to the adoption of several signal standards. They are affected by the power frequency employed, and the reproducible state of the art at the time the service was introduced. Thus, picture rates are 25 or 30 per second and line frequencies between 450 and 825 lines per second have been used. Now, most systems operate at 525 and 625 lines per second. As noted in Chapter 3, three basic techniques exist for encoding color: North America and Japan use NTSC; France, the Soviet Union, and Eastern Europe use SECAM; and most of the rest of the world uses PAL. (The NTSC color system is discussed in Section 3.4.1.) Digital processing is used to convert

convert video signals originating from equipment employing one standard to signals which can be reproduced on equipment employing another standard.

4.4.1. Television Receivers

The modern television receiver has been under development for the last 50 years. All major countries now provide television service and most have at least one color channel. In the United States, around 98% of all households have one television receiver. The penetration is slightly higher than telephone service (about 97%). Almost all of these households have a color receiver, and two-thirds have two or more receivers (black-and-white or color). Modern television receivers have all the characteristics of a consumer product. They are made to meet a price, to appeal to a market segment, to have features said to differentiate them from others and to work well—but not *too* well. The major technical component is the cathode-ray tube (CRT), on the face of which the color picture is formed by illuminating red, blue, and green phosphors formed in small dots or fine stripes. Theoretically, because of the discrete nature of this target and the finite width of the electron beam, picture resolution is limited to about 250,000 picture elements (approximately 600 wide and 400 high) in 525-line systems and to about 350,000 elements (approximately 700 wide and 500 high) in 625-line systems. In practice, a resolution of about one-half of these numbers is achieved. The resolution varies over the picture surface, being better close to the center of the tube, and worse close to the corners. Resolution is a parameter which has a major effect on the receiver. In the first place, it affects the detail observable in the picture and contributes to the picture *quality* perceived by the purchaser. Just as important, it is related to the angle over which the electron beam is deflected: the greater the angle, the more difficult it is to achieve good resolution in the corners of the screen, but the shorter the CRT will become, making a significant contribution to reducing the size of the receiver.

Suppliers of television programs have adapted to the resolution available. For the most part, the action and environmental components of entertainment and public affairs programs are adequately presented. However, text and graphics messages are not served so well. In fact, something less than one-half of a typewritten page can be displayed legibly by modern television receivers.

4.4.2. Digital Television

As in other electronics based industries, television system operators are replacing analog circuits by digital circuits to perform extended tasks better and cheaper. In Figure 4.28, the major elements of the studio, broadcast transmission, and receiver are shown. Analog techniques are employed in the broadcast transmission segment, in the sensors employed in the studio, and in the final reconstruction of the video message at the CRT and loudspeaker. Digital processing

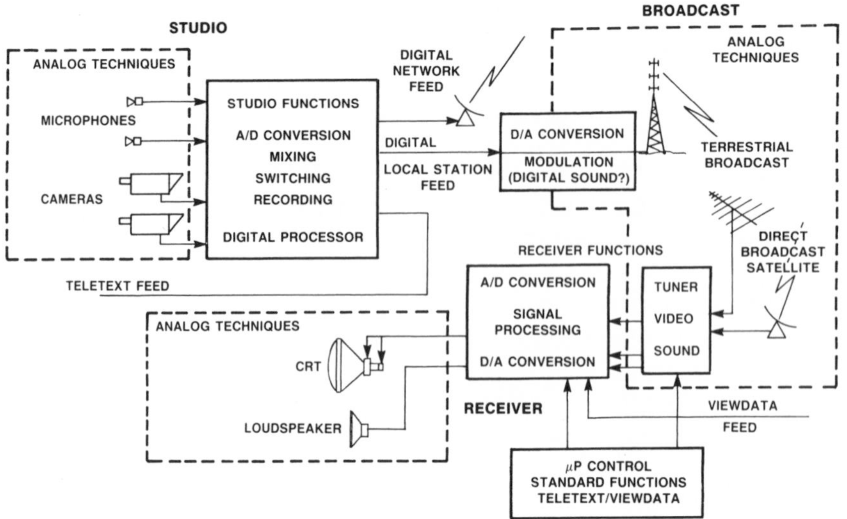

Figure 4.28. The television chain consists of studio and receiver functions which are implemented in digital fashion, and broadcast transmission which employs analog signals and modulation to conserve bandwidth.

and control are employed to provide studio and receiver functions. At the studio, mixing, switching, and other signal processing tasks are performed on digital signals derived from the analog signals provided by microphones and cameras. The composite signal may be distributed to other stations over a digital satellite channel. It is also fed to the local transmitting facilities where the analog signal is reconstructed for broadcast to the local audience in normal modulated format. At the receiver, this signal is demodulated into video, sound, and perhaps teletext components, which are then processed digitally to provide the information essential to forming the picture on the CRT and creating the sound in the loudspeaker(s).[19] Families of integrated circuit chips are available which: perform A/D and D/A conversion; separate luminance and chrominance and extract red, green, and blue components; set contrast, preserve color balance, etc.; control sound balance, tone, loudness, etc.; generate deflection signals which can correct for nonlinearities in the sweep to provide a degree of regularity impossible to achieve with only analog circuits; and control the tuner and other functions in a coordinated way, including remote and preprogrammed features. Through the increasing capability of VLSI, they will be reduced to a single chip.

4.4.3. Videotex Applications

The information retrieval systems known generically as videotex (see Section 2.3.1) use a color television receiver as the common display terminal. To make it suitable, special, mostly digital, circuits are added, as shown in Figure 4.29. Broadcast videotex (teletext) information is received on unused lines in the television signal. In the United States, the FCC has authorized the use of lines 14–18 and 20 of the vertical blanking interval. Line 21 is used for closed captioning applications. Wired videotex (viewdata) is received as a string of data characters over the telephone line.

Teletext presents special problems: the data rate must be fast enough so that a significant part of a page is contained on each signal line, the spectrum of the pulses used must fit within the allocated television channel bandwidth, and the pulse train must be receivable without significant modification of the receiving part of the receiver. Extensive tests have shown that a pulse rate of 5.72 Mb/s can satisfy these requirements for NTSC receivers when implemented with raised cosine pulses applied through a pulse shaping network at the transmitter which limits the pulse spectrum to 4.48 MHz. A pulse rate of 6.93 Mb/s is used in PAL and SECAM systems. Like standard television signals, teletext is susceptible to distortion due to echos caused by the delayed arrival at the receiving antenna of radiated energy reflected from tall buildings and/or other local obstacles. Unlike standard television signals for which echo delays over many microseconds are most annoying, teletext signals are rendered unintelligible by very short echo delays (<1.5 μs). Circuits which cancel these short echos are essential when operating in an environment where they occur. Teletext reception in areas of good television reception can be very satisfactory. However, because of the compromises made in band limiting the signal and the digital nature of the information displayed, there will always be fringe areas in which picture quality is satisfactory, but teletext is illegible.

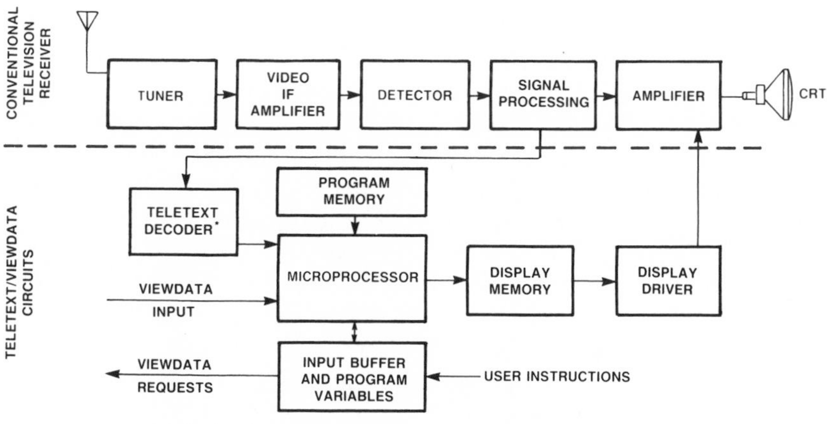

Figure 4.29. Videotex (teletext and viewdata) terminal formed by the addition of mostly digital circuitry to a conventional television receiver.

The videotex terminal shown in Figure 4.29 employs a microprocessor to receive and process the incoming data stream. Implicit in the diagram is the equality of data format from either system. While this need not be so, in those countries in which both broadcast and wired services exist, it is—for the very good reason that lower costs should ultimately be achieved with a common design. Input data words are interpreted using information stored in the program memory. The process depends on the display coding technique used. The result is stored in the display memory and used to construct a complete picture on the CRT screen. User requests for other pages of information, as well as initiation and termination instructions, are buffered and interpreted as viewdata requests or teletext requests in accordance with the options available to the user and the variations permitted by the service in use. In present practice two basic picture coding techniques are employed: alphamosaic and alphageometric.

Alphamosaic systems display characters or graphic elements in fixed cells positioned to form a rectangular grid. For graphics, each cell is subdivided into six subcells (2 wide × 3 high): they form 64 distinct mosaic patterns. A basic 7-bit code is used to provide 32 control instructions (carriage return, line feed, etc.) known as the C set and 96 symbol instructions (alphanumerics, punctuation, mosaics, etc.) known as the G set. For accented alphabets, an extended G set is necessary; for complicated displays, an extended C set is necessary. They are facilitated through the use of code extension techniques in which different meanings are given to a 7-bit word depending on what control word(s) precede it. In addition, a *dynamically redefinable character set* (DRCS) may be employed which provides additional (predefined) shapes that can be transmitted to the user's terminal as needed. Thus, a sequence of pages may each contain the same figure. With DRCS it can be transmitted to the user once, stored in memory, and invoked whenever a specific combination of C and G words is used. In principle, the flexibility of an alphamosaic system is limited only by the amount of memory included in the user's terminal. In actual fact, mosaic-based services use fixed format, character oriented systems, and are tied to the hardware limitations of a particular display technology. CCITT has recommended the adoption of a unified mosaic system which combines the separate implementations pioneered in the United Kingdom and France and includes DRCS technology.

Alphogeometric systems handle alphanumerics in the same way as alphamosaic systems. However, graphics are transmitted as a series of commands which define the construction of the desired shape. This mode requires more computational and memory capability in the user's terminal, but makes it possible to produce higher-quality pictures. In fact, the display quality will be limited by the precision with which the terminal is capable of responding to the picture description instructions (PDIs). PDIs are usually contained in an extended G set and invoked by designated control characters in the C set. For those images which require even greater detail, so-called "photographic" techniques can be used. They involve transmitting information picture element by picture element. Because of the large amount of memory required in the terminal, use of this technique is generally limited to a portion of the display, such as a signature block or roadmap. In principle, still color pictures of the same quality as a standard television picture can be produced. The technique is known as *alpha-photographic*.

The alphageometric/alphaphotographic techniques described above were developed in Canada and incorporated in Telidon. Since the announcement of the North American Standard Presentation Level Protocol for Videotex (PLP), Telidon has been modified to be identical to PLP (see Section 2.3.1.4.1). PLP incorporates all of the features of existing protocols but includes a very much wider range of colors and a broader range of features to make the display acceptable to commercial sponsors. Using the principle of the ISO open systems architecture model (see Section 3.10.2.1) the standard covers the formatting of data and is independent of the physical and logical implementation of the lower levels which are concerned with specific equipment. The standard includes [20]

- text characters identical to the earlier systems plus additional characters identical to NABTS (North American Broadcast Teletext Standard);
- DRCS procedures;
- block mosaics;
- geometric graphics using PDIs;
- C and G control sets.

A: VIEWDATA (WIRED VIDEOTEX) SYSTEMS

COUNTRY	DATA PRESENTATION STANDARD	PAGE FORMAT/ RESOLUTION	PAGE PRESENTATION RATE	DATA CODING	SERVICE NAME(S)
UNITED KINGDOM	TELETEXT	24 ROWS OF 40 CHARACTERS 6 × 10 DOT MATRIX	8 s/PAGE AT 1200 b/s	SERIAL ALPHAMOSAIC DRCS	PRESTEL
FRANCE	ANTIOPE	24 ROWS OF 40 CHARACTERS		PARALLEL ALPHAMOSAIC DRCS	TITAN, TELETEL
NORTH AMERICA	PRESENTATION LEVEL PROTOCOL (PLP)	240 × 320 PELS FOR TV RECEIVER 960 × 1280 PELS CAPABILITY		ALPHAMOSAIC DRCS ALPHAGEOMETRIC ALPHAPHOTOGRAPHIC	TELIDON
JAPAN	CAPTAIN	8 ROWS OF 15 KANJI CHAR. 24 × 16 DOT MATRIX	10 s/PAGE AT 3200 b/s	FACSIMILE	

B: TELETEXT (BROADCAST VIDEOTEX) SYSTEMS

COUNTRY	DATA PRESENTATION STANDARD	PAGE FORMAT AND PRESENTATION RATE	DATA CODING	SERVICE NAME(S)
UNITED KINGDOM	TELETEXT	24 ROWS OF 40 CHARACTERS 6 × 10 DOT MATRIX 24 s/100 PAGES	SERIAL ALPHAMOSAIC	CEEFAX ORACLE
FRANCE	ANTIOPE		PARALLEL ALPHAMOSAIC	DIDON
NORTH AMERICA	NABTS		ALPHAMOSAIC ALPHAGEOMETRIC	

NABTS = NORTH AMERICAN BROADCAST TELETEXT STANDARD

Figure 4.30. Characteristics of basic videotex systems.

The total of these capabilities will be adequate to provide high-quality displays for present information retrieval applications and to provide flexibility for future applications. In all likelihood a single videotex/teletext standard based on PLP will be adopted by Canadian and United States organizations.

In actuality, three standards are likely to emerge—one European, one Asian, and the other North American. In Europe, where 50 Hz, 625-line operation is normal, a page is likely to be 24 rows of 40 characters formatted using alpha-mosaic techniques augmented by DRCS. In North America, where 60 Hz, 525-line operation is normal, a page may be 20 rows of 40 characters formatted using PLP techniques. In Japan and those parts of Asia which employ ideographic writing, other standards will be used. Page format may vary from region to region. Between Europe and North America, compatibility will be asserted, and will be obtained through the use of complicated interfacing equipment. The characteristics of basic videotex systems are listed in Figure 4.30.

4.4.4. Other Approaches

Besides conventional systems which employ color CRTs and 525/625 line formats, techniques are being developed to provide higher-definition pictures and new types of displays.

4.4.4.1. High-Definition TV

With text and graphics likely to become more important in the future, proposals have been made for, and experiments conducted to demonstrate, higher-resolution television. Most systems incorporate 1000 to 1500 lines, employ a video (luminance) bandwidth of around 10 MHz, and a chrominance bandwidth of 5 to 7 MHz. At least one system has been proposed which is compatible with present standards. Using a bandwidth of 12 MHz, existing receivers would use the lower 6 MHz to create a standard resolution NTSC picture. High-resolution receivers would use the full 12 MHz signal to create a picture with significantly better resolution that approaches the limits noted in Section 4.4.1.[21] To avoid interference with existing facilities, some high-definition systems envisage broad-casting from geostationary satellites at 12 GHz. Many opportunities are afforded by a new system for changing existing parameters. For instance, picture size and aspect ratio can be expanded to give a wide screen display which can be projected on the wall for home entertainment, or can be reduced and rotated to make a direct reading screen that displays a full typewritten page. The number of scanning lines can be increased and the fraction which may be used for data can be adjusted.

For direct viewing CRT tube displays, brightness is adequate and the resolution achieved is determined by a combination of the number of scan lines, the line width, and the construction of the phosphor screen. For color, which requires a precise pattern of red, green, and blue phosphors, the phosphor spot size or linewidth must be reduced, and the number of dots or lines must be increased, as the number of scanning lines increases. For projection systems, the resolution achieved is usually limited by the brightness available. With contemporary picture sources, the light energy produced is less than sufficient to reproduce a 1000-line picture. For color systems which use three color projectors, high resolution requires an exact balance of the individual color tubes and requires more precise optics than can usually be included in commercial systems. For these reasons high-resolution systems favor direct viewing displays.

4.4.4.2. New Displays

At present, practical, direct-viewing, full-color, dynamic displays are provided exclusively on cathode ray tubes. Despite a large amount of research into alternative techniques, it remains the one component which can fulfill *all* requirements. In addition, any point on the screen is easily addressed by applying the appropriate voltages to the deflection coils and turning on the electron beam. Other display techniques are under investigation which seek to achieve the same performance. Results so far are not reassuring and the CRT is unlikely to be replaced in the near future. That is not to indicate that there will not be other techniques which will find specific applications. For instance, *liquid crystals* display the time and other information in modern electronic watches. Liquid crystals have also been used to build small, black-and-white television displays. Another technique uses the controlled, localized breakdown of a *gas plasma,* such as neon. Small, television-style displays have been made. They show the characteristic neon color. Yet another technique uses *electroluminescent material.* A voltage is applied across appropriate materials to cause them to luminesce. All three effects can be used to construct flat-panel displays which have specialized applications. A great drawback is the need to construct individual elements which represent each picture element. None of these techniques has the simplicity of sweeping a modulated electron beam across a three-color phosphor mosaic, and most are too slow to capture rapid motion.

4.4.5. Cable Television

Cable television systems transport multiple TV channels to a community of subscribers over coaxial cable connections. Through a drop cable each subscriber is connected to a feeder cable which carries a common signal from the head end, perhaps over an intermediate trunk.

4.4.5.1. Frequency Limits and Number of Channels

Early systems carried up to 12 channels at the VHF broadcast frequencies (3 channels from 56 to 72 MHz, 2 from 76 to 88 MHz, and 7 from 174 to 216 MHz) so as to be able to couple directly to the tuning capability of the subscriber's television set (channels 2 through 13). Improved amplifier and cable technology led to systems with a nominal 35-channel capacity operating from 50 to 300 MHz, and more recently, systems operating at 50 to 400 MHz to provide 52 channels. Five-hundred MHz systems may support 68 channels and 600 MHz systems some 85 channels. In some installations, two cables are employed, making over 100 channels available to each subscriber. Because the channel frequencies do not correspond to the tuning capability of standard television receivers, frequency conversion is performed in a separate unit, usually called a *set-top converter*. The frequency of the channel the viewer wishes to receive is converted to the frequency of an unused, off-air, VHF channel, to which the receiver is permanently tuned.

A CATV system is a wideband, analog system which transports a set of complex signals whose amplitude and phase must be faithfully preserved if the color picture is to be reproduced satisfactorily. The wider the bandwidth, the more difficult it is to achieve near perfect amplification. For one thing, amplifier power must increase as the number of channels increases in order to achieve an acceptable signal-to-noise ratio for each channel. Higher power and lower distortion are difficult to attain together. Because amplitude and phase distortions add from amplifier to amplifier, they limit the number which may be cascaded. To achieve satisfactory quality from a chain of around 25 amplifiers, the distortion level in each amplifier must be about −90 dB (the power attributable to the distortion products is 0.0000001% of the power due to a single TV channel at normal operating level). Because the attenuation of a coaxial cable increases by approximately 15% from 300 to 400 MHz, and by about the same factor from 400 to 500 MHz, the 300 MHz amplifier spacing of around 2.2 kft (660 m) will decrease by 15% and 30% at the higher frequencies. Stated another way, provided the cascade limit is not exceeded, the capacity of a 300 MHz, 35-channel system can be increased by almost 50% (from 35 channels to 52 channels) by increasing the number of amplifiers by 15% and increasing the operating frequency limit to 400 MHz. It is almost doubled by adding 30% more amplifiers and increasing the limit to 500 MHz.[22]

4.4.5.2. System Architecture

The main features of the topology of CATV systems are shown in Figure 4.31. In (A), the wideband, multichannel signal is shown transported on a feeder cable from which it is distributed on drop cables to individual subscribers. In (B), a substantial community is served by a single head end connected to the individual subscribers by a tree structure of trunks and feeders. In (C), a large subscriber population is supplied many channels through intermediate links which are connected by supertrunks to a central head end installation. The supertrunks may be provided by microwave radio, optical fibers, or low-loss coaxial cable.

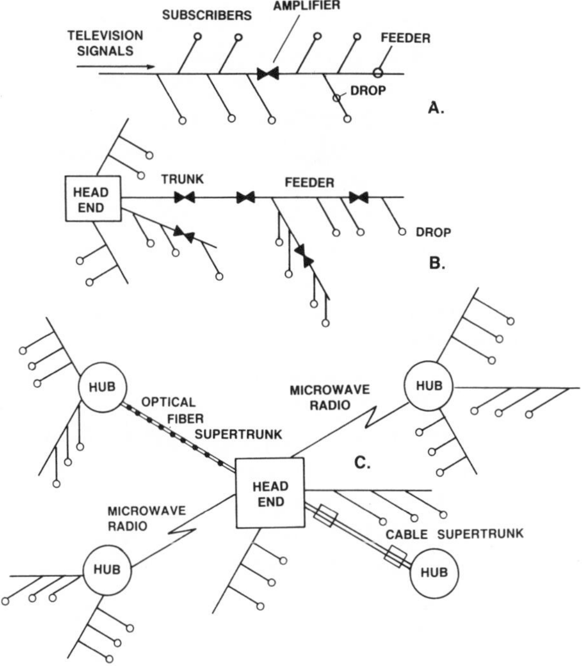

Figure 4.31. The main features of CATV systems. In (A) a wideband signal containing several TV channels is distributed to subscribers through drops from a single common feeder cable. In (B) a substantial subscriber population can be supplied with a modest number of channels by a tree structure of trunks and feeders connected to the head end. In (C) a large subscriber population can be supplied with many channels through intermediate links which are connected by super trunks to the head end.

At the head end, signals are collected from program sources through the use of

- antennas which capture off-air TV and intercept signals from satellites distributing national network programming, pay-TV programs, and special interest channels;
- microwave or other point-to-point links transporting signals from distant stations;
- signals originating in local facilities, and signals from video tape recorders and other equipment installed at the head end.

The signals are demodulated, conditioned, and some are encoded if they are to be used as programming for pay-TV, or are only for use by special interest groups. They are then frequency multiplexed to form the signal which is carried through the network to the subscriber. Many systems also provide a range of

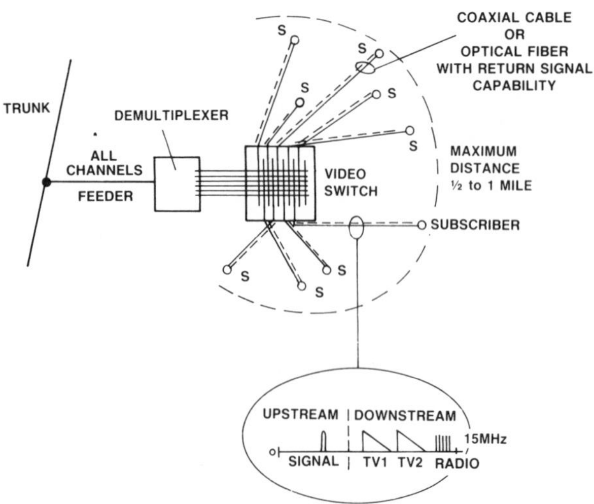

Figure 4.32. An alternative CATV distribution topology employs a switched system for the last few thousand feet. Only the signals ordered from those signals available to the subscriber are sent to the terminal.

audio signals derived from radio stations and other sources. In most systems, each subscriber receives all signals. Having all of the signals available to each subscriber reflects the historical development of CATV systems. Having all of the signals available allows some subscribers to *steal* signals by constructing their own unscrambling devices.

An alternative approach to the distribution of television signals is illustrated in Figure 4.32. It uses a wideband switch and star distribution topology for the final connections to a group of subscribers. Only those signals ordered by the user from the signals available to the subscriber are switched to the terminal. In this way the system operator can ensure that the subscriber is paying for what he uses. Implementation requires two-way communication. Downstream, the signal may consist of two or three TV channels, which can be selected from all the channels available on the system, and selected radio channels. In the upstream direction, signaling is required to select the programs the subscriber wishes to receive. So as to limit the expense and maintenance required, the distance from switch to subscriber must be less than the distance at which an amplifier would be required. The maximum distance from switch to subscriber is thus one-half to one mile. Under this condition, two-way communication can be provided over a single coaxial cable using FDM. Alternatively, the upstream and downstream channels can be separated by installing a twisted pair with each cable. The cable is used for downstream signals and the wire pair for narrow band upstream signals. Another option might employ optical fiber with WDM, or a twisted pair can be installed with each fiber. All are technically satisfactory solutions: which is employed in an actual situation depends on the premium to be paid for elegance. The provision of a return path opens up opportunities for the distribution plant to be used for low-bandwidth services associated with alarms, meter reading, polling, and like activities.

4.4.5.3. Two-Way Systems

Two-way communication may also be achieved over tree structured cable systems. Figure 4.33 shows the principle of a two-way system. Up to 52 channels are provided in the downstream direction in the frequency band from 50 to 400 MHz in the normal way. One or two channels are reserved for signaling from the head end to the household so that the system operator can

- provide information to the subscriber in response to a request or as the result of an administrative action;
- poll equipment in the household; and
- adjust the spectrum of programs available to the subscriber (by opening or closing taps external to the premises, or setting special circuits in the addressable terminal).

In the upstream direction, signals requesting service or information or those signals generated by alarms, meters, and in response to polling, are multiplexed in the frequency band 5 to 30 MHz, transported to the head end, demultiplexed, and used by the program control to adjust the programming available or are passed on to other service providers such as security or utility companies. The return bandwidth may be permanently allocated to each subscriber: one system uses 500 carriers spaced 20 kHz apart from 5 to 15 MHz to serve communities of subscribers. Besides data, such channels can carry voice. When matched with a voice channel in the downstream direction, this arrangement can *bypass* part, if not all, of the subscriber loop plant. Other systems use time-division multiplexing in which each subscriber is assigned specific time slots, or time slots are assigned at the time service is required (demand assignment). These systems can transport voice also. Yet other systems poll each subscriber in a group at regular intervals. When information is awaiting transmission it is transmitted as a burst before the system polls the next subscriber. Such systems cannot transport voice easily. Some systems provide return video channels: two carriers spaced 6 MHz apart can be located between 18 and 30 MHz and used for transmission of local video programming to the head end for distribution over the network. In order to achieve a distortion-free picture, it is essential that the input signal level be adjusted so that the signal strength at the first upstream amplifier is within the limits specified for linear operation. Thus the signal level at the transmitter must be adjusted in accordance with the cable distance to the first

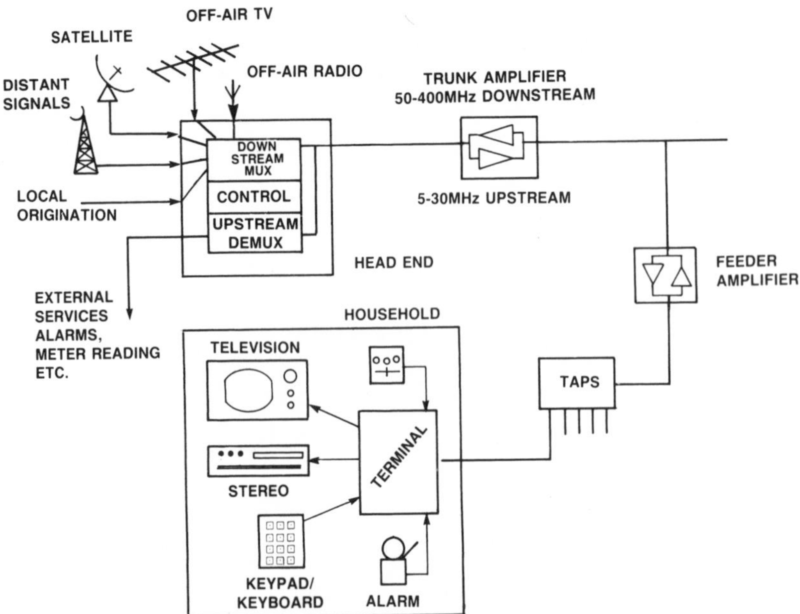

Figure 4.33. Principle of a two-way CATV system. Up to 52 television channels are provided in the downstream direction in the frequency band from 50 to 400 MHz. Access to all (or only some) is facilitated through the use of signal-controlled taps outside the household, or by addressable circuits in the terminal. In the upstream direction, signals requesting service or generated by alarms or meters from individual households are multiplexed in the frequency band 5 to 30 MHz, transported to the head end, demultiplexed, and used to adjust the programming available, or are passed on to other service providers. © 1981 IEEE.

amplifier. In fact, similar balancing must be done for all analog signals, and to some degree for all digital signals, as well.

4.4.5.4. Data Applications

Modern CATV systems have moved far from their simple beginnings as transporters of a few television signals to communities cut off from reception by terrain or distance. In fact, the largest facilities are now located in urban America where they serve not only millions of individual households, but also some of the needs of commerce, industry, and government. Penetration of the urban residence market is based on the ability to supply a wider variety of programming than the local broadcast stations can provide. In part, penetration of governmental, industrial, and commercial markets is due to the ability to supply television channels for institutional videoconferencing. However, an increasing amount of interest is centered on the ability of CATV systems to carry high-speed data, point-to-point. In many urban areas, CATV systems offer heavy data users the opportunity to transport data streams between facilities, or to earth stations. These connections are being made on coaxial cables and optical fibers.

4.4.6. Direct Broadcast Satellite (DBS) Television

Satellites are used routinely to transport TV signals from channel providers and superstations to *receive-only* earth stations which serve CATV systems. A few thousand 10–30-ft (3–10-m) diameter antenna dishes are in place around the country intercepting entertainment to fill up the 12 to 100 channels the cable systems provide. With the same antenna and frequency conversion equipment, private citizens may capture signals for private use. In the belief that television signals broadcast from space represent one more dimension which should be exploited to serve the communication needs of the public, FCC has authorized the construction of DBS-TV systems. The uplink will operate around 18 GHz in one design, and will carry three 16 MHz FM channels, each of which contains one TV program. Authorized broadcasting power will approach 200 W per TV channel at frequencies around 12 GHz. With this combination of high power and high frequency, the receiving antenna can be approximately 3 ft (1 m) in diameter. The front end of the customer's receiver will be mounted on the antenna and will down-convert the 12 GHz received signal to 1 or 2 GHz, a frequency which can be carried over relatively inexpensive cable to the TV receiver for further processing. The front end will likely employ submicron GaAs FETs or other devices in an integrated circuit structure. High-power DBS-TV is scheduled to start in 1986. Meanwhile, competing lower power systems using 20 to 40 W will be inaugurated in 1984. Eventually, high-definition television (HDTV) may be *spacecast*. Using in the neighborhood of 1100 lines, it will require the sort of bandwidths which can be allocated to broadcasting from space.

4.5. Radio

Radio broadcasting is a mature technology. Sudden changes in the method of implementation are not likely to occur. Nevertheless, changes have been made. For instance, the introduction of stereo FM broadcasting was accomplished by employing a technique which is compatible with existing monaural FM receivers. In approving the baseband spectrum for FM broadcasting (stereo and monaural), FCC made provision for the use of a sideband between 53 and 75 kHz to transmit programming of a broadcast nature to a limited segment of the public who wished to subscribe to these services. Called the SCA (Subsidiary Communications Authorization) channel, it has been used to provide subscription background music, reading services for the blind, and other specialized programs. The channel can also be used to broadcast data at rates up to 9600 b/s to support one-way teletextlike information services.[23] In April 1983, FCC extended the baseband from 75 to 99 kHz so that an FM station can offer two SCA services. At the same time, FCC permitted nonbroadcast applications, such as paging, electronic mail, data transmission, and facsimile transmission.

C. DATA COMMUNICATIONS

In this section, we describe facilities which are used for data connections. They consist of terminals and transmission switching equipment, some of which employ packet techniques.

4.6. Data Terminal Equipment

Access to, and control of, information is growing in importance in the modern world. Implementation of these functions requires terminals which map human communication modes into data processing commands and processed data into modes suited to human understanding. A conceptual model of a data terminal is shown in Figure 4.34. It imposes restrictions on the freedom of expression of the user and provides information in a limited number of formats. The need for *structure* makes data terminal equipment (DTE) different from telephone and

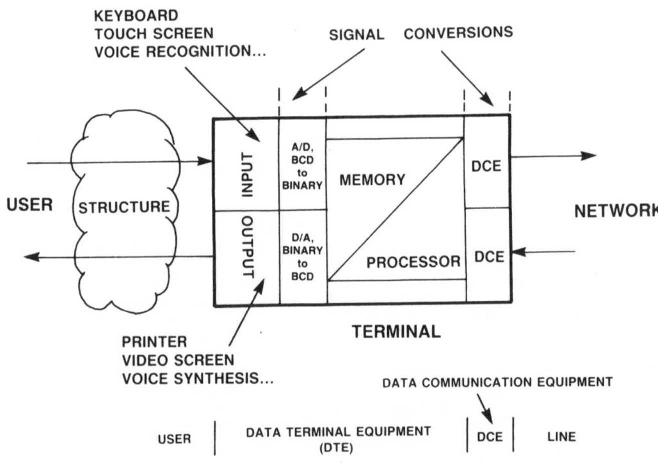

Figure 4.34. Conceptual model of a data terminal (DTE).

television terminals. It is a consequence of the interaction of man and machine and can give rise to feelings of frustration, anger, or fear in the user. Even though the data terminal can be no better than the combination of hardware and software which supports it, an important quality it must provide is to be *user friendly*. The sorts of input and output devices employed, and the amount and type of processing in the terminal, have an effect on this.

4.6.1. Input/Output Devices

The most visible portions of a data terminal are the *input* and *output* devices. Input may be accomplished through a keyboard which looks much like a standard typewriter. Commands and data are *typed* in a straightforward fashion using alphanumeric and special function keys. Each *keystroke*—the act of pushing a key down with a finger until a mechanism is actuated, then releasing the key and withdrawing the finger—is captured as a digitally coded signal which is used in the terminal and is sent to the data processing host. Movable keys and physical keystroking are being replaced by electronic panels and finger touching. On these keyboards, alphanumeric information is generated by touching keylike areas labeled in characteristic typewriter keyboard fashion. Operation relies on a change in capacitance or resistance effected by the finger contact. It can be constructed as a thin, smooth slab without moving parts and is amenable to touching, stroking, and wiping—finger motions which may be more *natural* than keystroking. For those who regularly use a DTE for professional or business pursuits, a separate keyboard appears to be an efficient command/data input device.

For casual users of DTE, other arrangements are possible. Thus, a keyboard may be displayed on a CRT on the face of which a thin plastic sandwich has been overlayed. The sandwich consists of thin, transparent conducting coatings separated by minute plastic projections. Touching the screen causes the conducting layers to touch creating a path for current to flow. By alternately measuring the flow between pairs of electrodes placed at the top and bottom, and the two sides, of the screen, the position of the finger can be determined. This can be related to the display area being touched and interpreted accordingly. Of course, the display on the screen need not be a typewriter keyboard but can be whatever combination of alphanumeric and special function areas are appropriate to the activity in progress.

Another technique uses buttons inserted in the terminal enclosure close to the edge of the screen whose functions are defined by labels displayed on the screen. The labels are changed as the user goes from task to task, eliminating the need to remember special key capabilities, or work with keyboards having a large number of keys.[24,25] Yet another simplification for the user is afforded by voice recognition electronics which can be trained to recognize individual word commands. For persons without keyboard aptitude, a combination of speak and touch is an effective way of addressing a data terminal.

Information can be returned from terminal to user in several ways. *Hard* copy from a printer, *soft* copy displayed on a video screen, and synthetic voice comments, are the modes listed in Figure 4.34. Both printer and video screen are likely to have been used to record the user's input so that a complete record of each information exchange is available. Searching the record is a matter of scanning the hard copy: with a video terminal, memory and logic must be provided so that the user can *scroll* through the electronic record.

4.6.2. Signal Conversion

No matter what devices are used to achieve the input/output (I/O) function, signal conversion will most likely be required between these devices and the remaining electronic equipment in the terminal. The most common transformations are analog-to-digital (A/D), binary-coded-decimal (BCD) to binary, and their complementary operations. They permit the core capability of the terminal to reside in programmable microprocessors and memories.

The degree to which a terminal resembles a stand-alone computer/data processor is a matter of user preference, application requirements, host support availability, and so on. As more user-friendly features are added, the basic operations will require more computer support resident in the terminal. As LSI chips evolve to VLSI and systems-on-a-chip, it will be easier (and cheaper) to provide significant processing power in the terminal. Many will assume the nature of communicating personal computers. Others will continue to be little more than communicating data entry/exit devices, needing the host machine and its databases to function.

Over the years, several codes have evolved for communication to and from terminals of the sort discussed. In North America the most common is the American Standard Code for Information Interchange (ASCII); in Europe and elsewhere, CCITT No. 5 code is common. Both are 8-bit codes in which 7 bits are used as characters, symbols, and control instructions (128 possibilities) and the eighth bit is used to check parity. Many of the entries in the two code sets are identical. If more than 128 bit patterns are required, the Extended Binary-Coded-Decimal Interchange Code (EBCDIC), can be employed. In this code, the eighth bit is used as part of the code to provide 256 specific patterns. Prior to transmission, the signal stream is converted to a form suited to the characteristics of the channel employed. Through the use of a modem, or data communications equipment (DCE), the binary signal is converted to, and reconstructed from, a form having a spectral distribution which does not overflow the bandwidth available.

4.6.3. Use of IR Links

For a changing office environment in which several DTEs must be connected to a central processing complex, IR (infrared) links can provide flexible connections. Infrared has the advantage over radio that it is blocked by solid objects so that the office walls isolate the space from the rest of the building. It has the disadvantage that the energy travels in straight lines—so that shadowing by large objects and reentrant corners can be an impediment to communication. Nevertheless, using diffuse IR, it is possible to construct an intraoffice system. Diffuse IR is achieved by radiating energy in a wide angle so that it strikes the walls and other objects and permeates the room. Thus, communication is achieved over several paths simultaneously and the impact of shadowing is reduced. This multipath sets the upper limit to the system bit rate: the theoretical limit is 260 Mb meters/s. For a room 13 m long, this corresponds to 20 Mb/s.[26] Other critical factors, such as the presence of ambient light and the digital modulation method employed, reduce the theoretical limit to a practical level of around 1 Mb/s. CSMA/CD techniques can be used to serve multiple terminals (see Section 4.7.4.1).

4.6.4. Facsimile

Facsimile machines which produce a replica of a document at a remote location have been available for many years. In the 1970s, representative units took 6 min to send a page which was reproduced with a resolution of around 100 lines/in. The document was scanned line by line to produce a signal proportional to the intensity (degree of whiteness or blackness) of each point on the line. This signal was used to modulate (FM) an audio carrier for transmission over analog lines. At the receiving end, the result was often a less than perfect copy due to phase shift and noise introduced in the transport network. Additional signal processing at the transmitter was able to correct some of the distortion, but only if the receiver was equipped and aligned in a complementary manner. A lack of compatibility between machines from different suppliers thwarted the growth of a national public service.

In the 1980s, representative units take a minute, or less, to send a page which is reproduced with a resolution of around 200 lines/in. Using digital techniques to represent each element, they employ processing to reduce redundancies in the signal. Most transmit at 4800 or 9600 bits/s and employ modems to access public telephone facilities. With 56-kb/s digital circuits, and employing line-by-line and line-to-line processing, machines can send a page in a few seconds with a resolution of 300 or 400 lines/in. With appropriate signal separation at the transmitter, and the use of three-color reproduction (using red, green, and blue inkjet printing, for instance) these machines can be used to send colored material.

The rapid pace of technical advancements is making standardization difficult. Nevertheless, CCITT has defined four classes of equipment. Groups 1 and 2 are 6- and 4-min analog machines, group 3 is for 1-min digital machines, and group 4 is for higher-speed digital machines. The latter will be able to support a near-perfect electronic mail system and will provide an impressive copying-at-a-distance (telecopying) capability.

4.7. Switching and Transmission

Data are transported and switched in public or private networks in voiceband channels which are part of existing voice facilities, as channels derived from existing digital facilities, and in packet networks.

4.7.1. Voiceband Data

To support DTEs operating over 4-kHz telephone channels, data rates of 1200, 2400, 4800, and 9600 b/s are employed. Frequency only, phase only, or phase and amplitude modulation schemes are used to produce efficiencies of 2 or 3 b/s/Hz. In Figure 4.34, DCE is shown integral with DTE. A more flexible arrangement is to separate them so that DTEs can be used with a variety of DCEs. Under this circumstance the DTE/DCE interface for telephone channel applications is defined in EIA Standard RS232C or CCITT Recommendation V.24 in terms of the connections to a 25-pin plug and socket. For higher performance DTEs, RS449 specifies the mechanical and functional characteristics of a 37-pin interface and RS422 and RS423 define the electrical circuit performance for balanced and unbalanced line connections. These standards are physical level protocols—Level 1 in the OSI model. For data rates us to 4800 b/s, most customer loops provide transport with acceptably low bit-error rates. A data rate of 9600 b/s often requires a private-line access loop which has been *conditioned* to eliminate any bridge taps and loading coils.

DTEs may communicate in both directions (i.e., receive and transmit) simultaneously, or in one direction at a time. Two-way simultaneous operation is known as *duplex* or *full-duplex:* one-way at a time operation is known as *half-duplex*. Duplex operation over a single wire pair is achieved through FDM, using carrier frequencies between 1000 and 1300 Hz for one direction and 2000 and

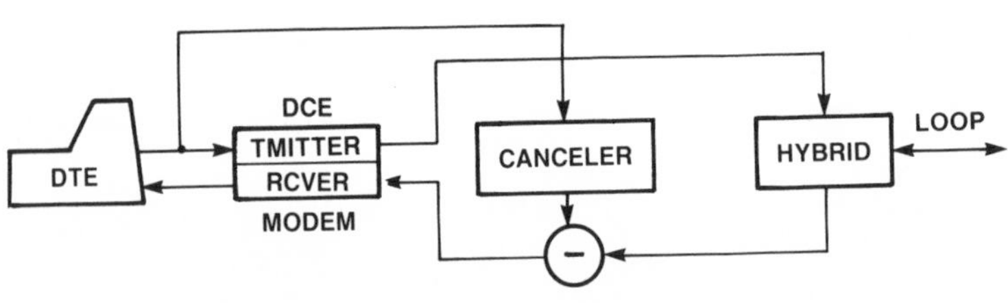

Figure 4.35. Principle of a digital data echo canceller used to achieve duplex working over a pair of wires (loop).

2300 Hz for the other. Duplex operation can also be achieved through the use of time compression multiplexing (TCM), a technique described in Section 4.3.1.3. Using line rates of between 150 and 300 kb/s, TCM can be used effectively over several kilometers.

Duplex operation using a digital echo canceller is also possible. Figure 4.35 shows the principle. Data from a DTE are connected through a modem (DCE) to a hybrid which transfers most of the signal to the wire loop. Some of the data signal returns to the receiver where it may be interpreted as a signal from the far end equipment. To prevent this, an electronic circuit is included which simulates the parameters of the leakage path and generates an inverse signal which just cancels the unwanted signal. It is set up by passing a burst of data around the circuit to allow the adaptive network within the canceller to achieve an initial balance. After start-up, a tracking algorithm adjusts the canceller when changes occur in the return path.

4.7.2. Circuit-Switched Digital Capability (CSDC)

CSDC uses the existing customer loop and requires special terminating equipment to support 56-kb/s data transmission. The principle is illustrated in Figure 4.36. A CSDC call is initiated by a voice call. Through the use of a

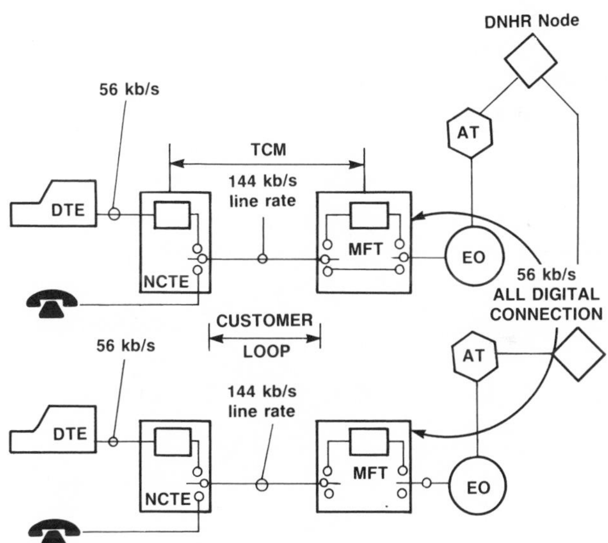

Figure 4.36. Time compression multiplex (TCM) is used on the customer loop to establish 56 kb/s circuit-switched digital capability (CSDC).

special prefix, the end office is alerted to complete the call over digital facilities and to make TCM equipment available. When the voice call is established and the DTEs are ready, the parties switch to data mode, and a 56-kb/s circuit connects the DTEs. On the two sections of customer loop the line rate is 144 kb/s as the DTE sends a burst in one direction and the end-office replies after a suitable interval. At this line rate the capability can be provided to over 90% of existing nonloaded customer loops. Although the majority of long-distance connections are analog (microwave radio and cable), a skeleton network of digital facilities has been created by adding a 1.54-Mb/s channel at the lower edge of the frequency bands assigned to the radios. Called *data-under-voice* (DUV), these channels are used for CSDC.

4.7.3. Local Area Data Transport (LADT) Capability

Another local arrangement is used to serve slow-speed, short-duration data traffic such as is generated by the use of viewdata and other interactive database services. The principle is shown in Figure 4.37. The customer at the top of the diagram has an either/or arrangement over which analog voice calls may be completed through the end office, or data may be sent to the statistical multiplexer (SM) by calling-up the appropriate port on the end-office switch. The customer at the bottom of the diagram has a special connection which uses a carrier to transport *data-above-voice* (DAV) directly to the SM. Here, the signals are separated, voice is passed on to the end-office switch, and data are passed on to the SM. The data incident on the SM are combined onto a 56-kb/s digital channel connected to a packet switch which routes the data to their final destination.

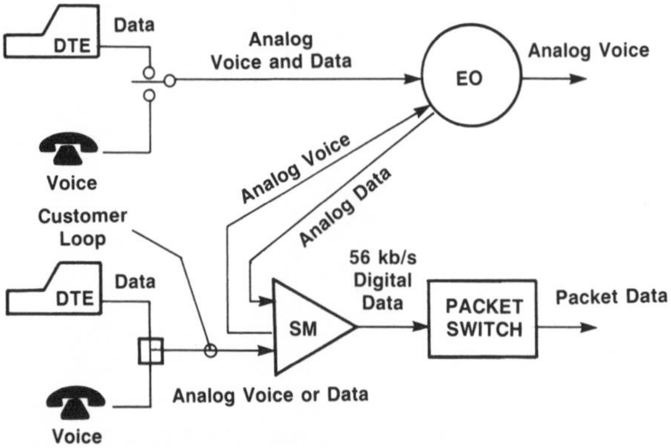

Figure 4.37. Analog voice and data are combined in the local area data transport (LADT) capability which connects voice to the end office and data to a packet switch.

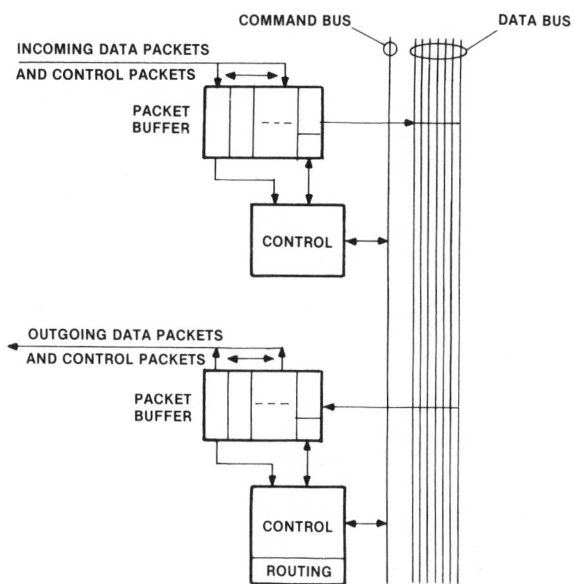

Figure 4.38. Principle of a packet switch.

4.7.4. Packet Data

The advent of data communications has encouraged the development of a switching technique suited to the *bursty* nature of the data messages. A packet switch accepts bursts of information (organized into packets of data), stores them in a buffer, and forwards them to the next switch in the packet network in accordance with the protocol employed. This technique is often called *store-and-forward* operation. Packets may be of fixed or variable length and contain thousands of bits. Besides message data, each packet has a header which contains source and destination information, and other parameters on which each packet switch makes decisions as to how to route packets, which packet to route next, whether to drop packets to reduce congestion, and whether to transmit packets.

4.7.4.1. Packet Switch

In a packet switch, call processing is significantly simpler than in a digital circuit switch since each packet contains the necessary routing information. Unlike a circuit switch, call setup and take-down are not required, nor must the controller maintain a map of busy connections (see Section 3.8.4). The principle of a packet switch is illustrated in Figure 4.38. It consists of a data bus on which packets are transferred in series, or in parallel. Often, the bus is made up of several parallel paths which allows the packet to be transferred byte-by-byte, or

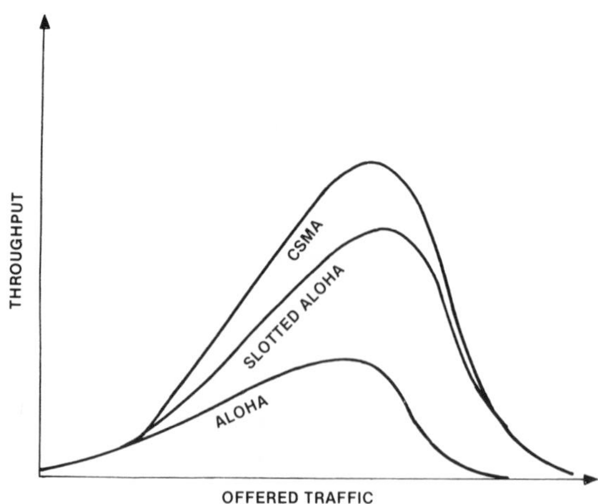

Figure 4.39. Throughput achieved by various packet-handling techniques.

even as an entire packet. Packets are transferred to and from the bus by processors which receive/transmit packets from/to other switches or terminals.

In Figure 4.38, incoming data and control packets are inserted in a buffer where they are checked for errors and then held ready for transfer to the data bus or interpreted by the controller to produce a control action. In the simplest case, transfers of packets to the bus occur at the convenience of the processor. If a collision occurs with a packet inserted at another port, neither reaches its destination port intact. Having determined that the packet is corrupted, the receiving port discards it. After a while, no acknowledgment having been received by the transmitting port, the packet is retransmitted at a time randomly selected by the transmit port control. Provided traffic on the bus does not amount to more than about 0.3 erlangs, there is a high probability of processing all incoming packets over the system. This mode of operation is known as ALOHA—the name of a system employing this technique which was pioneered at the University of Hawaii. As an alternative to random transfer, each port may be restricted to transmitting at the beginning of a timeslot (slotted-ALOHA) or may be required to sense that no other packet is present on the bus before transferring its data packet (Carrier Sense Multiple Access—CSMA). The throughput achieved by these alternatives is shown in Figure 4.39.

Other alternatives exist: CSMA/CD—a port may be required to detect that traffic is not present on the bus before transmitting, and to monitor during transmission to detect collisions (collision detection, CD); breakthrough—packets with priority are moved to the head of the input buffer and processed as soon as possible; priority routing—a port receiving priority packets may inhibit all others during transfer of its packets; TDMA—specific time slots may be assigned to each port mimicking the satellite operation discussed in Section 4.1.1.4.1. The regimen for these operating modes is established by messages passed on a separate command bus (or on channels on the data bus). The data packet is transferred from the bus to an outgoing buffer by that processor which serves the address the packet contains. Incoming and outgoing packets are buffered and released in accordance with commands from the receiving equipment.

4.7.4.2. *Protocols*

Providing the ability to interconnect disparate and dissimilar information processing users requires a family of standards and protocols which not only define the telecommunication facilities involved, but also standardize the functions performed by the terminating devices—terminals, computers, etc.—so that the information transported will be intelligible to the users. Organizing this family has been made easier by the development of the OSI model, described in Section 3.10.2.1, and the active work of CCITT based on this model.(27) The OSI Reference model applies to the interconnection of users requiring access to resources within the jurisdiction of at least one other user. The general problem is illustrated in Figure 4.40. *N* users with diverse equipment must be interconnected on demand through several public and private networks. The process of interconnection is complex. It includes some or all of the following: data code conversion, data structure mapping, transmission speed matching, flow control,

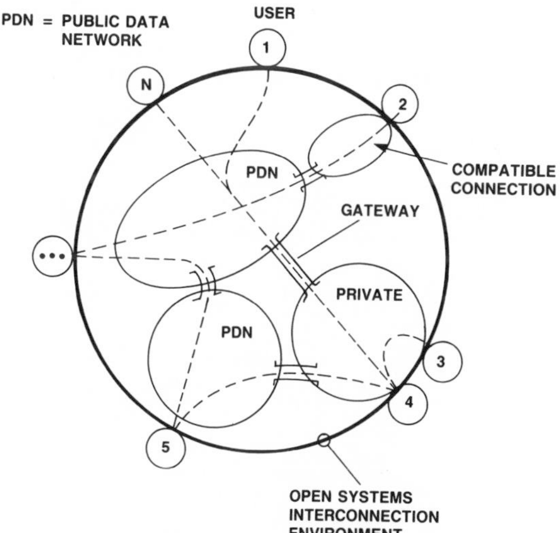

Figure 4.40. Illustration of the data telecommunication problem—interconnecting *N* users with diverse equipment through several public and private networks.

error detection and correction, routing, and encryption. As information passes from the originator to the recipient, each of these parameters must be selected appropriately, and may change as the traffic leaves one network and enters another.

4.7.4.2.1. CCITT Recommendation X.25. For packet networks an important protocol is X.25—CCITT Recommendation X.25—which makes a universal interface feasible between data terminal equipment (DTE) of varying degrees of sophistication and public packet switched networks. The interface consists of three parts: X.25 Physical Level, X.25 Link Level, and X.25 Packet Level. The physical level provides full-duplex synchronous data transmission while the link level formats the data into frames and uses cyclic redundancy codes to detect errors in the data stream transmitted by the packet level. Error recovery is achieved by retransmission of frames.

X.25 Physical Level incorporates the CCITT Recommendations to define the physical communications connection. X.20 and X.21 define the physical characteristics and call-control procedures for the DTE/DCTE (digital communications terminating equipment) interface for DTEs employing start–stop (X.20) and synchronous (X.21) transmission. Recommendations X.20bis and X.21bis are interim versions which cover the operation of asynchronous and synchronous modems on telephone networks. X.22 defines the interface between a DTE and DCTE operating at 48 kb/s and multiplexing a number of X.21 subscriber channels. Currently, all networks support X.21bis and data rates of 2.4, 4.8, and 9.6 kb/s. They also offer either 48 or 56 kb/s.

X.25 Link Level incorporates high-level data link control (HDLC) as a class of link access procedures (LAPs) known as the asynchronous balanced class of procedures, and designated LAPB. It allows DCTE to use the same repertoire of commands and responses as the DTE, making DTE/DCTE and DTE/DTE operation identical.

X.25 Packet Level defines the structure of the packet which includes both message data and administrative data. The format of a packet with HDLC framing is shown in Figure 4.41. The packet contains an address field, control field,

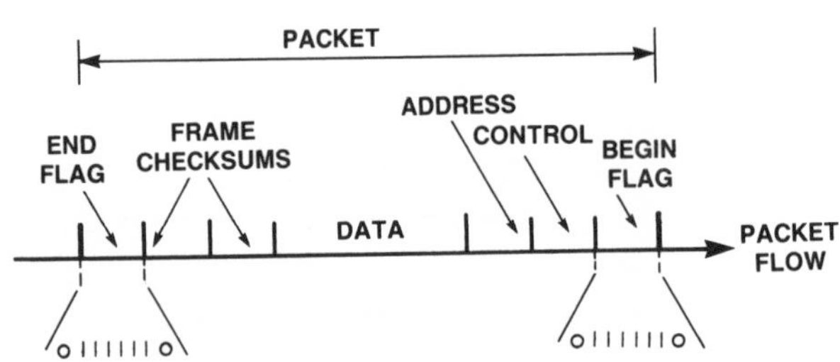

Figure 4.41. Basic packet structure includes fields assigned to flags, control, address, data, and checksum functions.

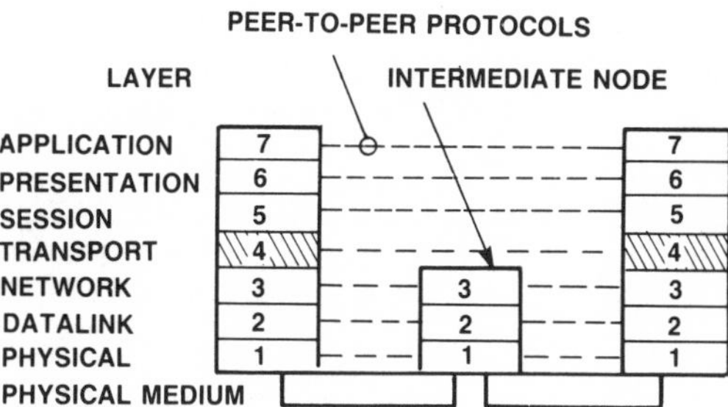

Figure 4.42. In a multinode connection, the intervening nodes operate below the transport layer of the Open Systems Interconnection (OSI) model.

message data, and frame checksums for error detection. Amongst other things, the control field is used in the following ways: to set up and take down the virtual circuit (in the first and last packets of a message only); to acknowledge receipt of packets; may instruct the sender to cease sending packets in order to control packet flow and prevent buffer overload; and provides packet sequence information. The packet is delineated by flags (01111110): to protect against the unintentional use of this sequence, the transmitting equipment inserts a zero into the data stream after a sequence of five ones. The receiving hardware deletes a zero which follows five ones.

4.7.4.2.2. Multinode Connections. The transport layer has overall responsibility for managing the flow of data between end systems in an open system interconnection environment (OSIE). It isolates the session and higher layers from any concern with the means of transportation and operates between systems which directly connect with the user DTEs. In a multinode connection, the intervening nodes operate below the transport level, as shown in Figure 4.42. The transport layer is responsible for meeting end-to-end quality of service requirements. Throughput, transit delay, error rate, security, and priority are some of the factors which must be managed in the context of end-system to end-system data transport.

4.7.4.3. Packet-Switched Networks

The architecture of a packet-switched network is shown in Figure 4.43. It consists of a fully connected backbone of Class 1 nodes, supported by a hierarchy of Class 2 and Class 3 nodes. The major transmission connections are 56-kb/s or 1.54-Mb/s circuits. The Network Control Center monitors delay in the network, intervenes to provide alternative routing as required, and collects accounting and operating information. Two basic modes of network operation can be distinguished: datagram and virtual circuit. In *datagram* mode, no attempt is

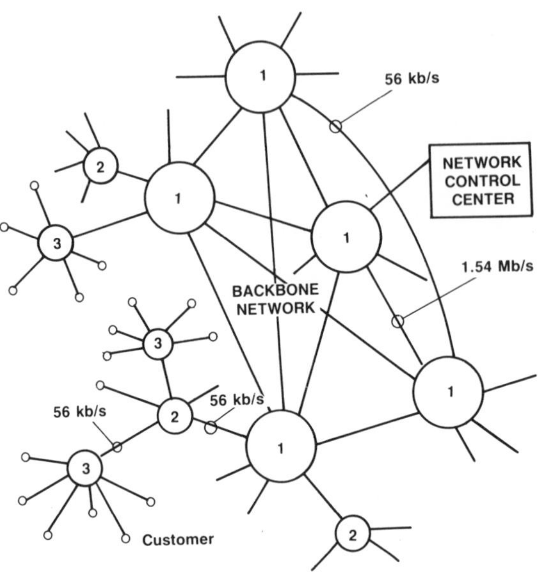

Figure 4.43. Representation of the architecture of a packet-switched network.

made to deliver packets in the same sequence as they entered the network—they transit along any available path, and are delivered in whatever sequence they arrive at the receiving terminal. In *virtual circuit* mode, the order of delivery matches the order of entry. This is achieved by using a pathfinder packet to establish a logical path through the network which is followed by all packets in the sequence. Even though they arrive in sequence, the time interval between these packets will vary depending on the amount of traffic presented at each node and to each transmission link. A measure of network activity is provided by the number of packets handled each second, by each node. Values of over 1000 packets/s/node may be experienced for peak-hour traffic.

4.7.4.3.1. Congestion. If user demands exceed the system capacity, congestion may set in rapidly. Figure 4.44 compares the performance of an uncontrolled network with the ideal. At a certain level of offered load, overload begins, and packets are not delivered on the first attempt due to the occurrence of full buffers at points in the network. With further increases in offered load, the uncontrolled network reaches maximum throughput, then degrades rapidly as congestion builds up until all buffers are filled with administrative messages (acknowledge, repeat, receiver not ready, etc.), and deadlock ensues. The function of the flow control features embedded in the protocol layers (i.e., link, network, and transport levels) is to prevent this catastrophe. The result is the smooth curve in Figure 4.44. This behavior is achieved by increasing the number of packets which are not delivered on the first attempt.

4.7.4.3.2. CCITT Recommendation X.75. Connections between networks are usually referred to as *gateways*. The interworking of networks offering packet-switched service is the subject of CCITT Recommendation X.75. It is consistent with X.25 and is supported by several other recommendations concerned with signaling, transit control, etc.

4.7.4.3.3. DARPA Internet. The Defense Advanced Research Projects Agency (DARPA) of the Department of Defense (DoD), continues in the forefront of the development and operation of packet-switched networks. From the original ARPANET (see Section 3.10.2.3), the pioneering packet network of the early 1970s, this agency has continued to support work directed to solving the operational problems of ever-larger networks. Today, the DARPA Internet interconnects diverse types of packet networks in the United States and Europe linking from 400 to 500 host computers. Gateways between the networks are provided by computers which *home* on single packet switches located in two (or more) individual networks. When a packet is received by a gateway, the input network header is modified to conform to the requirements of the transit network, routing information is expanded or modified, and the packet is dispatched to the next gateway, or its destination. A gateway-to-gateway protocol determines connectivity to both networks and neighboring gateways and implements a dynamic, shortest path, routing algorithm. Internet supports remote host and file access and electronic mail services on a regular basis. Services such as packet voice, facsimile, video graphics, multimedia mail, and teleconferencing are being developed and tested.[28]

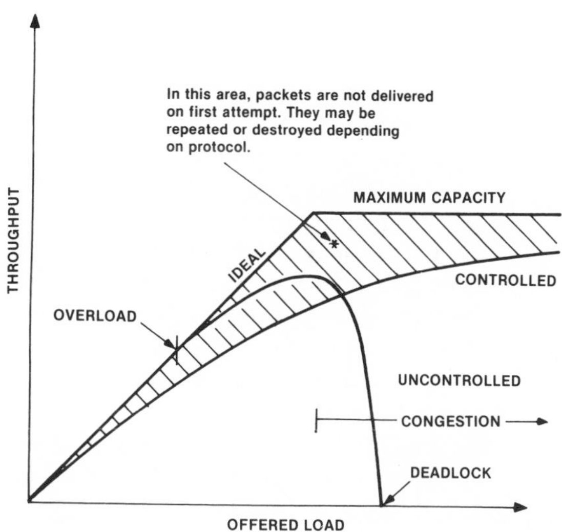

Figure 4.44. Flow performance of ideal, uncontrolled, and controlled packet networks.

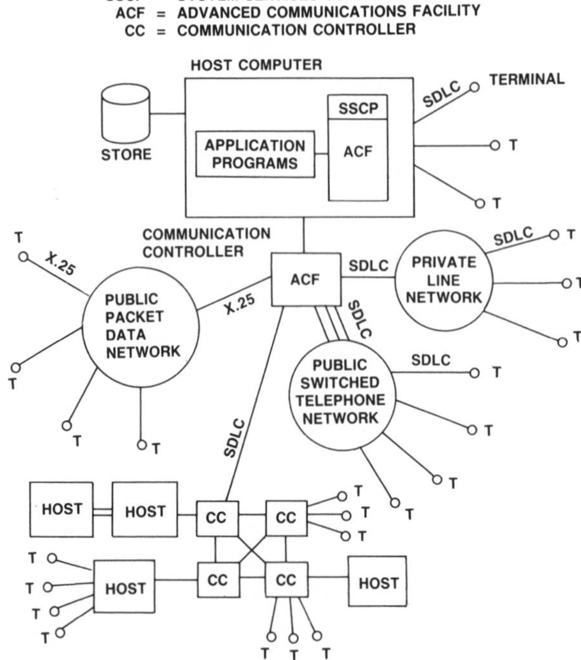

Figure 4.45. An SNA network.

4.7.5. SNA Networks

SNA networks generally consist of many terminals accessing a single host computer, or a small number of hosts. Local terminals are connected directly to the host computer. Distant terminals may be connected over private lines, through the public switched telephone network, or through public packet data networks, to a communication controller which connects to the host. In turn, this controller may connect through other communication controllers to additional host computers and terminals. The general arrangement of an SNA network is shown in Figure 4.45. Access from host to host and terminal to host is permitted over multiple routes.

In an SNA network responsibility for the layered protocols (see Section 3.10.2.2) rests with the advanced communication facility (ACF) which is resident in the communication controller and the host computer. Messages are divided into strings of frames which are passed from node to node using SDLC (synchronous data link control) protocol, or X.25 protocol for those nodes connected over a packet network. Message queuing is restricted at intermediate nodes, and throughput is controlled by pacing the rate at which frames are transmitted. An interface program, backed up by buffer storage, is required to incorporate X.25 working in the network.

4.7.6. Digital Transmission Service (DTS)

DTS employs omnidirectional, microwave radio equipment (operating around 10.6 GHz) to provide 1.54 Mb/s channels which support two-way communication between a user and a local node at distances up to 6 miles (10 km). In a metropolitan area, service may be provided in several such areas. To prevent frequency interference, each area is divided into sectors (or cells) by using 90° or 120° antennas and assigning an appropriate fraction (one-quarter or one-third) of the transmit and receive frequencies to each sector. A further isolation can be achieved by employing alternating horizontal (H) and vertical (V) polarizations. A three-sector, frequency reuse pattern is shown in Figure 4.46. Transmissions from the local node to users may employ continuous-broadcast *time-division multiplex* (TDM). Each user extracts his information from the time slots assigned to him. Transmission from the user to the local node may be made in time slots assigned to the user at the time of transmission (demand assigned TDMA). Other protocols may also be employed. The arrangements for nodal access are shown in Figure 4.47. Additional frequencies are reserved for digital radio links from local nodes to a central hub (city node) and for intercity con-

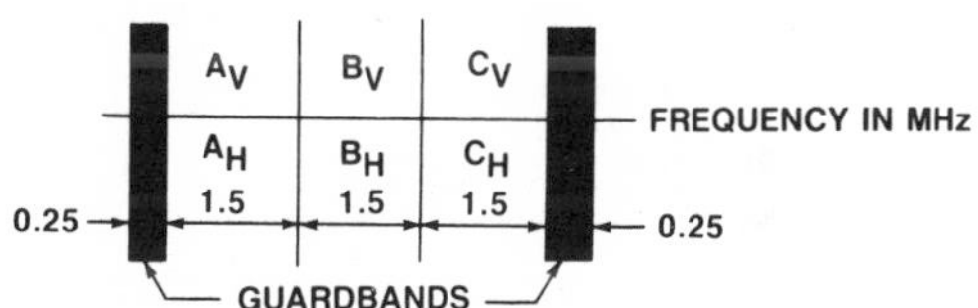

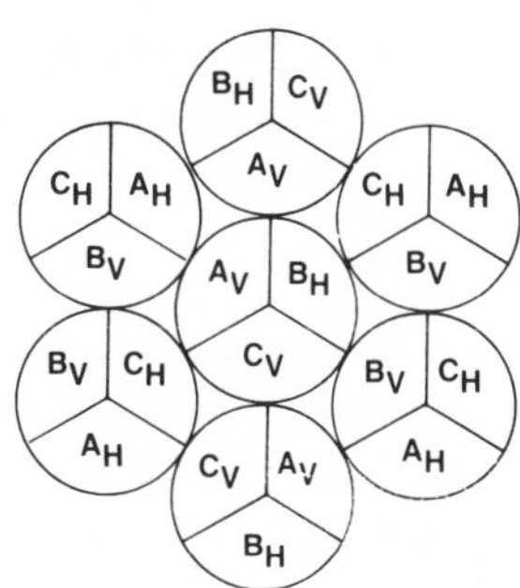

Figure 4.46. Frequency reuse pattern for digital termination service (DTS) employing three frequencies and two polarizations to cover seven cells.

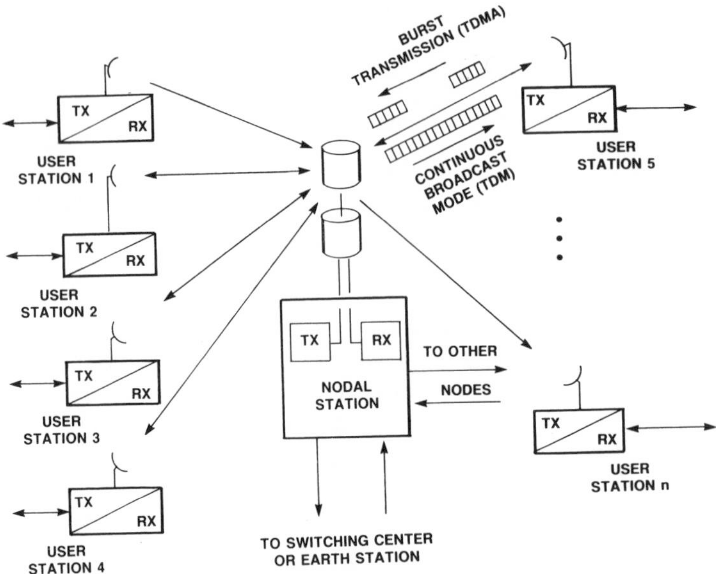

Figure 4.47. Digital Transmission Service (DTS) nodal access configuration provides digital radio channels between a local node and a community of users over distances of several miles.

nections to form *digital electronic message service networks* (DEMSNETs). The FCC has allocated four channel pairs (10 MHz/pair, i.e., 5 MHz transmit and 5 MHz receive) to extended DEMSNETs (networks serving more than 30 cities), and six channel pairs (5 MHz/pair) to limited DEMSNETs (networks serving less than 30 cities).

Spread-spectrum techniques (see Section 3.3.5) may be used for DTS channels. A direct-sequence spread-spectrum system transmits and receives a signal which includes a spreading code known to both the transmitter and receiver. The code occupies a much wider bandwidth than the information carried. Error-free demodulation requires that the receiver must be synchronized with the transmitted code. Because the receiver does not initially know exactly where in the code sequence the transmitter will start, a period of time is required for the receiver to acquire, synchronize, and track the transmitted code.

4.8. Local Area Networks

Within an enterprise, the interconnection of DTEs, processors, and other related systems to achieve the transmission, reception, exchange, and manipulation of information can be provided by a *local area network* (LAN). While it can be accomplished in several ways, LAN is usually taken to mean a common packet communications backbone which links work stations and other equipment (acquired from multiple vendors) on demand.

4.8.1. Network Architecture

Ring, bus, and star structures can form the architectural basis for LANs. They are sketched in Figure 4.48. In example A, each station is connected to two adjacent stations. Data from one of them are received by the middle station and passed to the next station. All of the stations in the network are linked in this way to form a closed loop, or a *ring*. When a station receives a packet of information, it is examined and, if addressed to the station, moved to a working storage for use. If the packet is not intended for the station, it is passed on to the next station. Should the station wish to send information, the packet is assembled and introduced into the ring at a suitable moment. It is received,

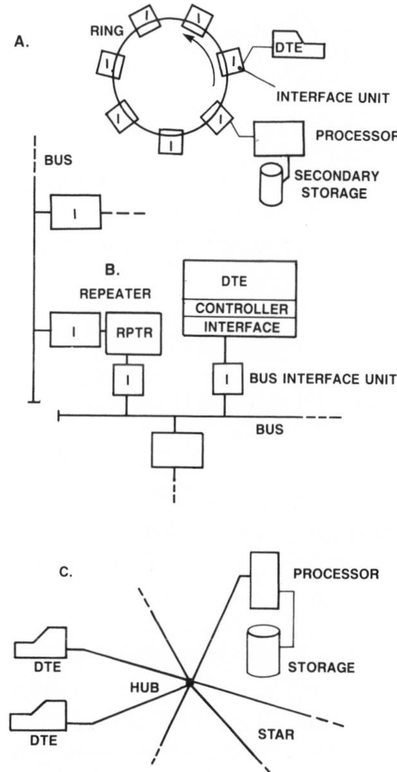

Figure 4.48. Basic ring, bus, and star architectures for LANs.

examined, and retransmitted by each station until it arrives at its destination. In a variation of this scheme, the station receiving a packet addressed to it copies the packet for use, marks the original packet received, and transmits it on around the loop. When the marked packet returns to the originating station, it is removed and the fact that it was received is noted. If there is no acknowledging mark, the packet will be retransmitted by the originating station. Because each station receives and transmits all of the information on the ring, it follows that all stations must be working correctly for the ring to function. In practical systems, redundant connections, connections to stations next-but-one away, and provisions to connect input to output to eliminate a failed station may be included to ensure that the remainder of the ring continues to operate in the event of the failure of one node.

In Example B in Figure 4.48, stations are connected to a common bus so that each is connected to all other stations and immediately receives every packet inserted on the bus. Busses can be interconnected as shown. The repeater transfers packets intended for stations on the other bus. Because each station is only connected to the bus, and is not part of the common communication highway, all stations need not be working for the system to function.

In Example C in Figure 4.48, all stations are connected to a hub which transmits the signals received to all branches. It may be a passive device which simply interconnects all lines, an active device which repeats the input signals, or an active device which administers the network, resolves conflicts, imposes priorities, and generally takes on the characteristics of a packet switch. Because each station is only connected to the hub, all stations need not be working for the system to function. However, the hub is vital to operation and must be many times more reliable than individual stations.

4.8.2. Transmission Media

In existing systems, the physical transmission medium is generally coaxial cable. It can be used in all configurations. Fiber optic cable has attracted considerable interest. Its superior bandwidth and freedom from electromagnetic interference make the technology attractive. Because optical T-connections are expensive and lossy, fiber is not the medium of choice for bus systems (coaxial cable is). It can be used in both ring and star configurations. In a ring in which messages are passed in one direction only, the fiber connects point-to-point between the stations. Connections to the next-but-one station which are invoked if the adjacent station fails can also be made with point-to-point fibers—or a controllable tap can be installed at the adjacent station. What to do becomes an interesting study in the probabilities of failure of the various configurations. In some rings, messages circulate in both directions. A message is assigned to the direction which provides the shortest route to the destination (the other direction becomes the alternative route). In this case, each direction of transmission can be served by a single fiber, or a single fiber can support both directions of transmission through the use of optical multiplexing (see Section 3.12.5). The same is true of a star configuration: two-way transmission can be achieved in a single fiber, or two fibers can be employed. In any particular situation the configuration will depend upon the cost of the additional components required to use a single fiber, and the cost of the additional fiber required to use two fibers. They must be weighed against the possible failure modes and the reliability required.

4.8.3. Protocols

As with other packet systems, a protocol must be employed which provides adequate throughput without loss of packets. CSMA/CD is frequently used for bus structured systems, and can be used with star configurations. It cannot be used with ring structures. Another protocol is known as *token passing*. It can be employed with all systems. In this scheme, permission to transmit is passed to all stations in sequence. If the station receiving permission (the token) has nothing to transmit, it passes the token along to the next station. If there is a packet to be transmitted, transmission takes place before the token is given up. Token passing ensures that each station is allocated the opportunity to transmit, and eliminates collisions. But what happens if the station holding the token should suddenly fail? The network will cease to function unless provision is made to generate another token after a reasonable interval. And how are new stations added to the token passing sequence? Network control must include provision for this. Despite these complications, token passing has been proven to be a robust, efficient, contention-resolving protocol suited to LAN operation.

Studies of the comparative advantages of token and CSMA/CD protocols have been made which take account of the offered load, the message length, and the topology. The results suggest that[(29)]

- on a lightly loaded bus, the use of CSMA/CD produces shorter delays than the use of tokens;
- CSMA/CD throughput performance is better the shorter the bus, and the longer the packet;
- a ring which employs token passing can carry more traffic than a bus which employs token passing.

For star configurations, CSMA/CD and token passing protocols can be used. With passive hubs, it is likely that throughput performance depends upon the length of the individual connections and packet length. With active hubs, it seems likely that throughput will increase in proportion to the processing power they contain.

5

Integration

The modern office requires voice for conversational purposes, data for information retrieval, and video for teleconferencing. In increasing numbers, voice, data, and video channels are being used by the same customer. Separate facilities to switch and distribute each type of message can be expensive, and may not be necessary. With the pervasive application of digital technology comes the possibility of integrating much of the switching and distribution system, thereby saving facility space and expense, making coordination of services simpler, and providing synergistic combinations leading to enhanced service opportunities. A result of this sort of integration is the need to replace familiar terms such as telephony—which implies a star-connected voice network—with others such as point-to-point, point-to-multipoint, and broadcast (one-to-many) facilities, which describe the connection and do not preclude the use of combinations of different bit-rate (and bandwidth) channels.

5.1. Integration of Point-to-Point Voice and Data

Long-distance transport needs will be supplied by analog radio and coaxial cable for some time to come (see Section 4.1.1.1.1). However, there will be a steady growth in digital facilities. Optical fiber and digital radio will encroach on the analog domain as economics permit and digital switches require. The continuing decreases in costs and increases in capabilities of digital electronics guarantee that digital techniques will eventually dominate point-to-point telecommunication. Even today, new starts are predominantly digital and there will be a time when renewing analog installations with more analog equipment is less attractive than conversion to digital to match the offered traffic. Competition in interexchange transport will ensure that these facilities are driven more by the first cost of new equipment than the book value of in-place equipment, and the price of a digital channel may be set at a level to attract traffic and achieve economies of scale with which to vindicate the price structure.

Meanwhile, exchange area facilities are experiencing a rapid conversion to digital equipment as in-place T-carrier systems are supplemented or replaced by optical fiber and digital radio, digital end offices are installed, digital line carrier systems are placed in the loop, and preparations are made to support digital telephones and other digital devices on the customer's premises. The growth of digital applications and the penetration of digital techniques to the full range of point-to-point telecommunication facilities will encourage the handling of voice and data services over the same facilities. In fact, the facilities may also handle that point-to-point video which is associated with teleconferencing and which is reduced to a digital stream of 3 Mb/s or 1.54 Mb/s or even 56 kb/s.

5.1.1. Integrated Services Digital Network (ISDN)

ISDN, which seeks to provide integrated voice and data (and digital video) services, is a network concept first formalized by CCITT in 1972 (Recommendation G.702). It envisaged an integrated digital network (IDN) to be used to establish connections for different services. (IDN is a network in which connections established by digital switches are used for the transmission of digital signals.) In 1980, in Recommendation G.705, ISDN was defined as evolving from IDN by incorporating functions and features to provide for new and existing services which are compatible with 64-kb/s switched digital connections.

The transition from existing networks to IDN and thence ISDN is recognized by CCITT to require a period of time extending to 10 or 20 years during which interim arrangements must be made for the interworking of services on ISDNs and other networks. When implemented, ISDN will provide digital channels which can be used simultaneously for voice and nonvoice services. They will display end-to-end digital connectivity, the users will have a common access arrangement for all services, and the channels will share terminal, transmission, and switching facilities (to the degree possible). The ultimate aim is to provide a digital *pipeline* to each user of whatever capacity the telecommunication needs may require.

ISDN may employ a number of bit rates to support a format which includes digital voice, digital data, and digital signaling. For instance, an information rate of 80 kb/s provides a 64-kb/s voice channel (B channel), an 8-kb/s data channel (B' channel), and an 8-kb/s signaling channel (D channel). If duplex transmission in the customer loop is achieved through the use of electronic hybrids and echo cancellers (see Section 4.7.1), an additional 8 kb/s may be added for synchronization. If duplex transmission is achieved through time compression multiplex (TCM, see Section 4.3.1.3), bursts of 22 bits at 250-μs intervals and a line rate of 256 kb/s may be employed. Other information rates are being

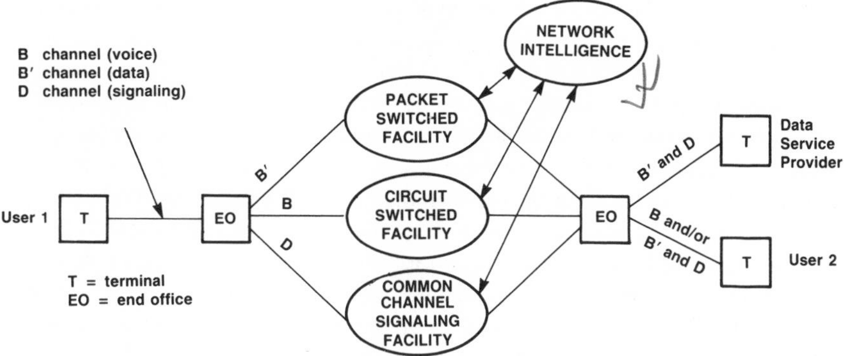

Figure 5.1. Illustration of the ISDN concept. User 1 is connected to user 2 over a combination of facilities which support voice (B channel), data (B′ channel), and signaling (D channel) messages. User 1 may also be connected to a data service provider using only B′ and D channels. Coordination of the simultaneous use of voice and data channels will require the use of additional network intelligence.

considered. At 144 kb/s, for instance, $B = 64$ kb/s, $B' = 64$ kb/s,and $D = 16$ kb/s, providing a higher-speed data channel, and more signaling capability.

Figure 5.1 illustrates the concept of ISDN using functionally separate facilities for the transportation of voice, data, and signaling messages. If a packet-switched connection is not available, data will be carried over circuit-switched facilities. If a common channel signaling connection is not available, signaling will be accomplished over the circuit-switched facility or over the packet-switched facility (if it is present). The simultaneous use of voice and data channels will require control and routing intelligence over and above that which is normally associated with the individual facilities. This capability is referred to as *network* intelligence. It is generally considered to consist of processors which coordinate routing and multiple channel use, collect traffic statistics on multichannel use for routing and facility planning purposes, coordinate billing records, and support new multichannel user services as they are defined. Network intelligence will be widely distributed throughout the system and will require an architecture of its own to ensure the proper interworking of the modules.[(1)]

ISDN emphasizes the compatibility between new services and switched 64-kb/s digital connections. In this context, a 1.54-Mb/s stream becomes 24 × 64 kb/s channels which can be handled in parallel in digital end offices to provide a switched videoconferencing service (for instance). An important requirement is that the switch must process all channels so that the parallelism of the streams is not disturbed. The bits arriving together must be switched and leave together. Modern digital end offices can be programmed to do this. ISDN should also accommodate speech encoded at 32, 16, and 8 kb/s so as to be able to take advantage of the benefits of digital chip technology. How to do this is one of the matters for future study by CCITT.

Despite Recommendation G.705, the evolution to ISDN is far from smooth. The world's telephone administrations are proceeding to introduce digital technology at different speeds, and there is no common view of the needs for future services—particularly which services to support first. Complicating the issue further is the variation from country to country of those who have the right to provide telecommunication services.[2] But perhaps the greatest obstacle is that the pace of digital development is so rapid that there may not be enough time to plan and implement an orderly evolution to ISDN.

Whether or not a master plan exists, where economic and legal conditions permit, services will be introduced which build on existing facilities to meet user demand. In the United States, for instance, both CSDC (circuit-switched digital capability) and LADT (local area data transport) capability, discussed in Sections 4.7.2 and 4.7.3, support voice, and/or data services. In CSDC, the circuit is time-shared by analog voice and digital data, one mode at a time, and signaling is provided over the analog voice channel. In LADT, circuits may be time-shared between analog voice and analog data, one at a time, or analog data may be carried above analog voice in parallel on the same two-wire connection. Largely, facility sharing is limited to the customer loop and end office. Both CSDC and LADT are pragmatic solutions to connecting voice and data customers to the network, now!

5.1.2. Voice and Data Local Area Network

Voice and data can be carried together over a local area network. In Section 4.8, a local area network (LAN) was defined as a common packet communications backbone which links work stations and other equipment on demand. These networks can be used for a combination of data and voice traffic by converting the digital voice signal to a series of packets. Special handling is required.

5.1.2.1. Packet Voice

Voice packets are formed from digital voice streams by placing a number of speech samples in each packet. To achieve faithful reproduction, the packets must arrive at their destination in a timely fashion, in the order that they were generated, so that the contents can be read out in a continuous 64-kb/s stream to form 8000 amplitude-modulated pulses per second from which the exact frequencies of the original speech can be reconstructed. Minor variations in packet arrival time can be compensated for by the use of buffer storage at the receiver. Larger variations will result in the loss of speech samples because they are not present to be used when required. The arrival of a packet out of sequence will distort the sound. For these reasons, voice packets should transit a network in a virtual circuit (so that all packets in the stream traverse the same route). In addition, to prevent the out-of-sequence arrival of a duplicate packet for one which is lost, those elements of the packet protocol which require the transmitter to repeat the packet if receipt is not acknowledged must be suppressed (see Section 3.10.2.1.1). Further, at switches where packets contend for resources, voice packets should receive priority.

The quality of packet voice depends on the number of lost packets and the length of the packets. For short packets (10 × 8-bit speech samples) a loss of more than 1%–2% of the packets is noticeable. For longer packets (100 × 8-bit speech samples) a smaller loss (1%) becomes noticeable. A significant improvement can be achieved by filling the gap produced by the lost packet by repeating the last packet received, or by repeating the tail of this packet several times. With compensation of this sort, around 2% of packets of either size can be lost before their absence is noticed.[(3)]

Packetizing a continuous 64-kb/s voice channel produces packets which contain at least one byte which is not zero (corresponding to periods when the speaker is speaking) and empty packets which contain all zeros (corresponding to periods when the speaker is not speaking). By dropping the empty packets, packet voice streams can be multiplexed together to achieve channel gain. Using packets short enough to take advantage of silences within a talkspurt results in channel gains of three or more. A practical limit is reached when the packet overhead (flags, control, address, and checksum bits) becomes a significant fraction of the message.

5.1.2.2. CSMA/CD Voice and Data Network

Local area networks which employ carrier sense, multiple access, collision detection (CSMA/CD) protocols (see Section 4.7.4.1) can transport voice and data simultaneously. The number of voice circuits which can be handled depends on the delay allowed, the permissible number of lost packets, and the speed of the bus. A 1-km (0.6-mile) bus, operating at 1 Mb/s can support 7 simultaneous conversations with short packets, and 13 simultaneous conversations with long packets. Assuming a call blocking probability of 0.01 and an activity of 0.25 erlang per telephone, the network can support from 10 to 26 telephones. A 1-km, 10-Mb/s bus can support 90 simultaneous conversations with long packets, enough to support 280 telephones.[(3)] Processing and buffering delays can be many milliseconds. For conversations between digital telephones on the LAN, the two directions of communication are separate so that echo is not a problem. Echo control will be required should the conversation occur over the LAN and be completed by facilities which contain two-wire terminations.

In a simultaneous voice and data application, to the extent that the data packets cannot be interleaved between the voice packets, data traffic will reduce the number of telephones which can be supported. For a system in which the data ports are used for slow-speed (<2.4 kb/s) data terminals, the number of data ports which can be active will be very much greater than the number of telephones.

5.1.3. Voice and Data PBX

PBX features have evolved to support both voice and data communication. Figure 5.2 shows the general configuration of a modern PBX. While something less than 100 calling features may be included in any installation, a generic capability of as many as 250 voice features and 150 data features is often available. Some of the more popular voice features are: automatic line selection on handset pickup, distinctive ringing (outside vs. inside calls), line status display (idle, busy, ringing, and holding), conference calling, call transfer, hold with reminder tone, automatic dialing (last number dialed, saved number dialed, abbreviated dialing), privacy, and call restrictions. Because of the complexity of accessing so many features, intelligent telephone terminals are available which contain microprocessors to sort through the protocols required to implement them. Voice and data messages derived from telephones, data terminal equipment

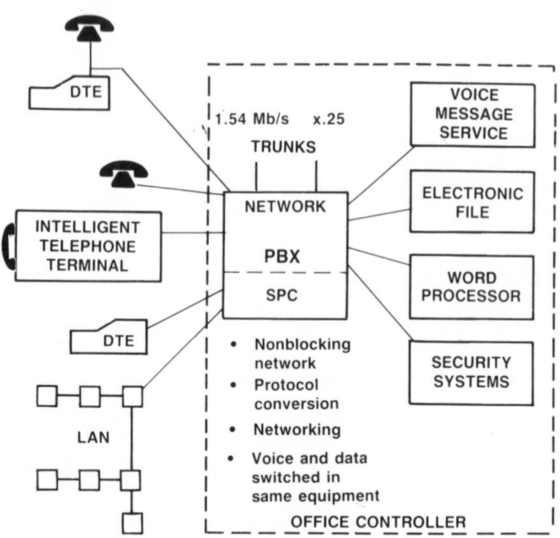

Figure 5.2. A modern PBX switches voice and data derived from a range of sources in the same equipment. Some systems incorporate services such as voice messaging, electronic filing, word processing, etc. They may be called office controllers.

(DTE), and LANs are switched in the same equipment under stored program control (SPC). The network is essentially nonblocking and connects to 1.54-Mb/s trunks or packet connections. Special interface circuits may be incorporated which provide protocol and speed (bit-rate) conversion so that terminals manufactured to different requirements can communicate.

Modern PBXs can incorporate features which permit interconnection so as to form private networks. Features such as uniform network numbering, message detail recording, automatic route selection, remote administration and maintenance, and remote testing allow the implementation of systems which can carry all of the internal telecommunication of geographically distributed corporations. End-to-end digital connections can be obtained through the facilities of inter-exchange, exchange area, or bypass carriers. Besides person-and-person, terminal-and-machine, or machine-and-machine connections, some systems include voice messaging, electronic filing, and word processing. They represent a distributed computer-based system which might be better called an office controller.

5.1.4. PBX or LAN?

As we have learned, both PBXs and LANs can be used to switch voice and data. Neither of them is *best* for all situations. LANs can support small to medium populations of voice and large numbers of data terminals, but bus speeds are not adequate to handle large populations of telephones. In an environment in which there are few telephones and many data terminals, a LAN is preferred. In an environment in which voice requirements are medium or large, the bits associated with voice overwhelm the bits associated wtih data and make a PBX the preferred equipment (irrespective of the number of data terminals). PBX and LAN can be used in combination (as indicated in Figure 5.2). The LAN is used to interconnect the bulk of the data terminals and to connect to one of the ports of the PBX which switches the voice circuits and provides external communication connections for terminals which need it. If the 64-kb/s channel speed is a limitation, it can be overcome by switching channels in parallel. If very high speed data communication is required (tens of Mb/s), it can only be achieved over a LAN.[(4)]

5.1.5. Integrated Voice and Data Switching

The integration of voice and data facilities will change existing traffic patterns due to the mixing of voice and data calls. Each hour, many thousand of credit authorization inquiries are directed toward urban databanks—each call requires only a few seconds. Other transaction services associated with banking, brokerage, and like enterprises may require seconds, not minutes. Connections used for teleconferencing will be engaged for an hour, or so, and connections to a time-sharing computer service may last all-day. For public network switches

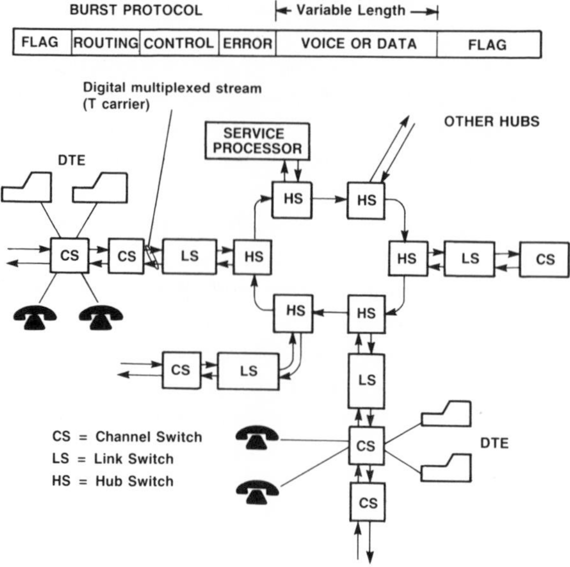

Figure 5.3. Principle of a burst switch.

which have been engineered to serve a customer community characterized by a three-minute voice call, these new holding times present problems. Calls of a few seconds each make greater use per unit time of the shared equipment which sets up the connections. Calls of several hours duration make greater use of specific switch connections and may make path selection for other calls more difficult. The result is a decline in the grade of service to which customers have become accustomed.

To circumvent most of these effects requires a switching technique in which routing and call information is carried along with the message rather than being transferred to functions in the switch. It calls for a technique which employs packets, yet one in which voice delay and distortion are prevented. A concept now in development shows promise of doing this. Called *burst* switching, it maximizes the ratio of information carried and transmission capacity utilized by handling both voice and data as bursty signals. Routing and control information are carried along in the messages. The general arrangement of a burst and a burst switch are shown in Figure 5.3. A burst consists of a bit stream of variable length divided into a header containing routing and control data with error protection, a burst of voice or data, and a terminating flag. Voice bursts are variable in length and contain entire talkspurts. Data bursts are of packet length and are packed between the voice signals. An important feature is that burst switching elements are distributed throughout the network. These independent, decentralized, digital switching units provide opportunities to improve the productivity of outside plant.

A user is connected to the burst switching network through a channel switch (CS) which provides access to a digital channel in a multiplexed digital stream (such as a T carrier). Switching is based on the information contained in the burst header, and the logical connection lasts for the entire burst. So as to give priority to voice bursts, the CS may delay data bursts (by buffering them). The mixed voice and data burst streams are passed through several CSs to a link switch (LS) where the bursts are prioritized and prepared for insertion in a ring of hub switches (HSs). Each HS passes information to one neighbor. A burst is circulated until it arrives at the HS which connects through a link switch to the channel switch serving its destination. The link-and-ring structure is sized to be nonblocking on the basis of the capacities of the transmission facilities and the channel switches. No common control or store-and-forward functions are needed in the intermediate switches to control path establishment and information transfer. The hub switch is required to operate at very high speeds (around 100 Mb/s, for instance) which are compatible with contemporary silicon bipolar ICs. In fact, the architecture lends itself to implementation on a set of chips each of which implements a CS, LS, or HS.

If the called party is connected to the same CS, or a CS in the same link, bursts are routed directly by the CS. If the called party is on another link, a routing request message is sent to the service processor which returns the information for insertion in the header for call setup. If the destination is not available to the hub switch, the routing request is forwarded to the next higher hub, or an adjacent level translation and routing processor. Routing information is always provided back to the calling interface for call propagation. Call establishment consists of sending messages between source and destination to establish a duplex *virtual* connection. Call disconnect is accomplished by a message which releases the virtual circuit path.[5]

The system is contention based—it is possible for there to be more sources requesting service than there are channels available to serve them. Under this circumstance, three actions are possible: data are buffered, freeing up channels for voice and control information; data and all but priority control messages are buffered; and, as a last resort, speech samples are discarded, a technique known as *freezeout*. Freezeout is manifest as clipping of the front end of the voice burst data stream and affects speech quality. On a given link, it is approximately uniform for all sources and can be diminished by deliberately delaying the burst. This may affect overall speech quality adversely.

5.1.6. Reduced Bit-Rate Voice

Eight-thousand 8-bit samples per second was established as the standard for digital voice by the telecommunication organizations of the world in order that voice quality would not suffer unduly even after many conversions from analog to digital and back to analog again. In the period of transition from an analog network to ISDN, coding and decoding might take place as many as half-a-dozen times in a long-distance call carried on facilities of several administrations. For this reason, ISDN emphasizes 64 kb/s as the basic digital stream in the network and all facilities employ this rate. As more digital facilities are introduced, the need for so robust a coding will diminish and the number of bits in a sample can be reduced. Indeed, coding schemes other than PCM can be employed, and the voice bit rate reduced to 32, 16, or even 8 kb/s, (as described in Section 3.2.2). In transmission applications with enough channels to take advantage of silence detection, statistical multiplexing will provide an additional effective reduction of two. The penalty for the bit rate reduction is delay and echo. (However, future integrated circuit terminals can be expected to have echo cancellation built in, and when all facilities are digital, there will be no echo source.)

When all facilities are digital, the technique used to encode voice can be quite different from today's 64-kb/s companded, PCM. While no one expects this situation to develop soon in public networks, digital private networks can take immediate advantage of the newer techniques to reduce bit capacity requirements. The economic advantages of reduced bit rate coding are likely to put strong pressure on administrations to modify their 64-kb/s standard. The operational effects will ameliorate the dominance of voice in mixed service applications and may result in the wider use of LANs and other packet-based arrangements.

5.1.7. Reduced Bit-Rate Video

In Section 3.4.2, the use of reduced bit-rate video for videoconference purposes was discussed. Bit rates of 3 and 1.54 Mb/s are used and developments as low as 56 kb/s show great promise. Assuming that an acceptable solution is forthcoming at 56 kb/s, videoconference signals can be handled in the same manner as voice signals in an ISDN.

5.2. Business Services

Facilities which support data processing, information retrieval, electronic message distribution, teleconferencing in all its forms, and voice communications are being incorporated into contemporary offices. Insofar as near instantaneous access to, and exchange of, information speeds up operations and substitutes for other forms of communications (such as travel, mail, etc.), it can improve productivity. For organizations which operate in business areas in which market shares and sales margins are established by competition, this opportunity to reduce expenses can be irresistible. The result is that a new telecommunications service which has an expectation of benefitting the business community and the potential of a reasonable return on investment will have innovators ready to support it as soon as facilities are in place to provide it. Those services which continue to demonstrate a return will survive to generate revenues for both providers and users. Two ways to reduce the incremental expense associated with new services are to share basic equipment and make use of multipurpose terminals.

5.2.1. The Automated Office

A representation of the telecommunication and related data processing facilities and services which may be present in the business environment is given in Figure 5.4. The top layer is concerned with the categories of users and the uses to which the services are put, the middle layer contains terminating devices, and the bottom layer includes the local and long-distance connections of whatever bit rate (or bandwidth) are required to transport the messages offered by these devices. The connections are made through a stored-program-controlled PBX, and over a local area network. Person-and-person activities usually take place over telephones, television, or related equipment for the purpose of conversations or conferencing. Person-and-machine activities require the use of data terminals in order to insert, manipulate, and retrieve information for record and other

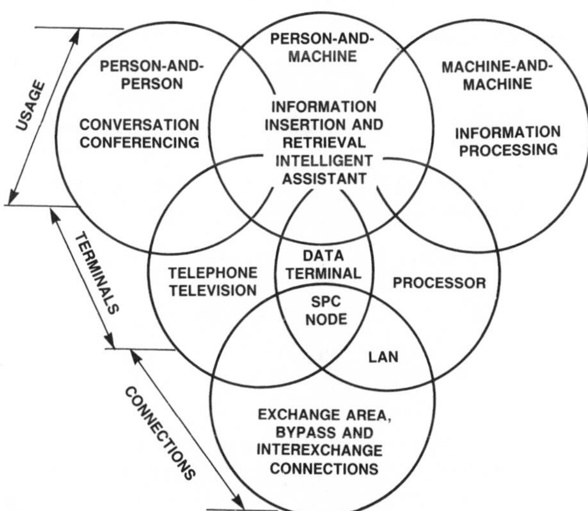

Figure 5.4. Representation of telecommunication and data processing facilities and services in the business environment.

purposes, and may be aided by intelligent assistants (expert systems) to complete complex tasks requiring judgment and experience. Machine-and-machine activities are usually an integral part of information processing operations. The diversity of uses, terminals, and connections presents both difficulties and opportunities.

The use of technology to improve the effectiveness of persons performing office-based tasks has come to be known as office automation. An *automated* office is likely to incorporate data processing services such as word processing, facsimile, electronic files, etc., to achieve desk-to-desk transfer of text, data, and image information as easily as voice information is now transferred over the telephone; computer services to accomplish the design and drafting of complex products and to control and support manufacturing; administrative programs which automate the preparation and coordination of calendars, provide reminders, and schedule activities; and communications services to provide electronic mail, information retrieval, and teleconferencing. Automated offices employ the established and emerging telecommunications media—and significant amounts of data processing and computer support besides—integrated into an assortment of communication arrangements.

The communications elements of an automated office which facilitate the generating, handling, storing, retrieving, utilizing, and communicating of information are shown in Figure 5.5. They consist of diverse terminals (including integrated work stations); message, information, and storage subsystems; LANs and a PBX; and communications over local and long-distance connections. In part, the communications requirements will be voice connections between individual work stations in the office and between work stations in the office and others outside. In part, the communications requirements will be data connections between individual work stations and the local computer, between workstations and remote computers, and between local computers and remote computers. In part, the communications requirements may be video connections between a limited number of local and remote terminals for videoconferencing, and from a central location to a large number of local terminals for information and training purposes.

Within each office, office building, and office park, signals may be carried on twisted pair, on coaxial cable, or on optical fibers, in star, tree, ring, or bus arrangements, or they may be broadcast using electromagnetic or optical energy. Configurations will depend on the bandwidths to be provided, the functions to be performed, and the distances to be covered, as well as the degree to which future computing and communicating equipment is merged. Signals may be analog or digital and the same medium may employ both. Thus, analog voice will serve standard telephones, although it will likely be converted to 64 kb/s to transit the PBX; and digital voice may be incorporated in integrated work stations which may transmit voice and data as a multiplexed, ISDN format stream, as packets, or as bursts. Because of the high cost of analog-to-digital conversion

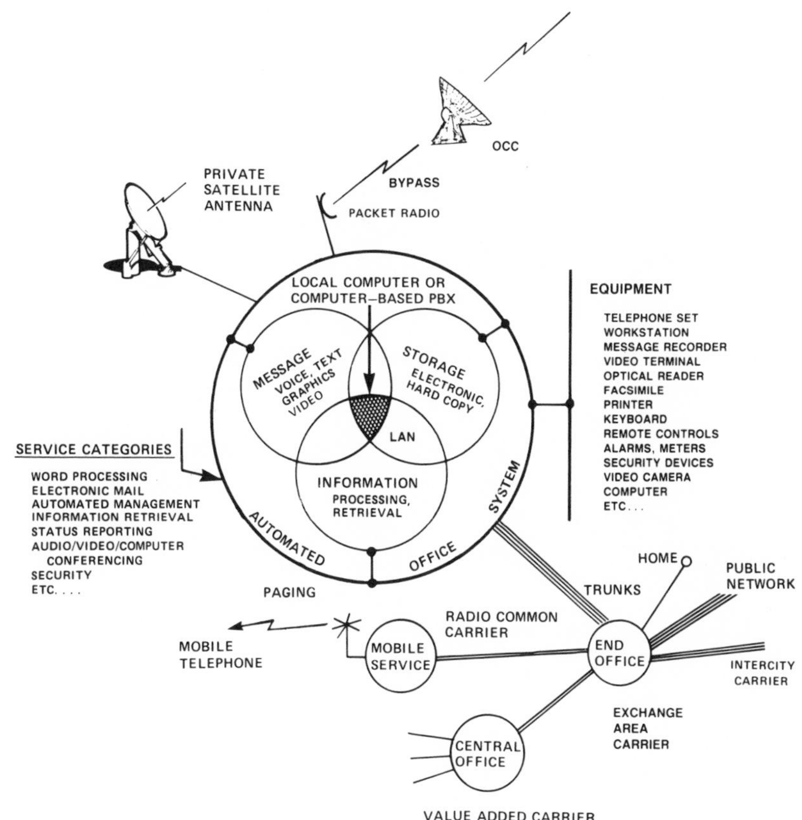

Figure 5.5. The elements of an automated office. Communications within the office facilitate the application of advanced technology to improving the productivity functions. They can be provided on twisted pair, coaxial cable, optical fibers, or use electromagnetic and infrared radiation. Communications outside the office may be supplied by several carriers. © 1981 IEEE.

(and vice versa), intraoffice video signals are likely to be analog: they will be converted to reduced bit rate digital for long-distance transport.

The specific functions performed in an automated office depend on the capabilities of the equipment installed. They may be stand-alone, or communicating units. By providing the opportunity to exchange information electronically (by voice, data, graphics, and video), the latter adds dramatically to the tasks which can be accomplished. The ability to reach a far-off databank makes central storage feasible and ensures that all of the available data can be assembled and reviewed at one time. The opportunity to exchange ideas and discuss problems with distant experts provides a higher level of expertise everywhere. By

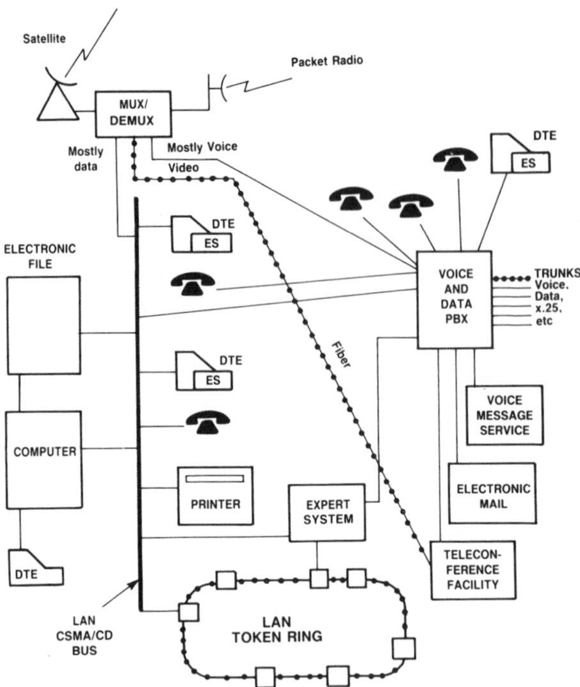

Figure 5.6. Configuration for the telecommunication facilities in an automated office.

fostering these actions, telecommunication merges the individual offices of the geographically distributed corporation into an extended electronic workspace in which limited media are employed in a close to real-time environment to attend to the details of the enterprise. Automated office is changing the way in which people transact and manage business. Further, it is changing the operating skills required by clerical and management personnel and is affecting the opportunities for personal interaction and satisfaction. However, the lack of equipment standards and of user-friendly, multifunctional software is impeding progress towards an all-embracing, office information utility. In the next ten or twenty years, such difficulties can be overcome through the use of individual integrated circuit chips which will provide each user with a custom-tailored, expert system (see Section 3.9) to ease the person and machine interface and make it possible for sophisticated, but not specially trained, users to work with information and other resources without knowledge of its whereabouts or the means by which it is obtained.

Figure 5.6 suggests a configuration for the telecommunication facilities in an automated office. Telephones and data terminal equipment (DTE) are connected to a voice and data PBX, as well as a CSMA/CD bus LAN supporting data and some voice services, and a token ring LAN supporting data services. Connections to the outside world are completed over trunks supplied by the exchange area carrier, local loop bypass facilities (such as packet radio and CATV), and an earth station providing connection to a satellite operated by an interexchange carrier. Some of the internal and external connections are made over optical fibers. Expert systems (ES) are included in some terminals, and as shared equipment available to other users, so as to facilitate the use of the entire range of system capabilities by the average worker.

5.2.2. Telecommunication Connections

As indicated in Figure 5.7, telecommunication connections will be provided by exchange area facilities, over bypass facilities, over private facilities, and over interexchange or packet network facilities.

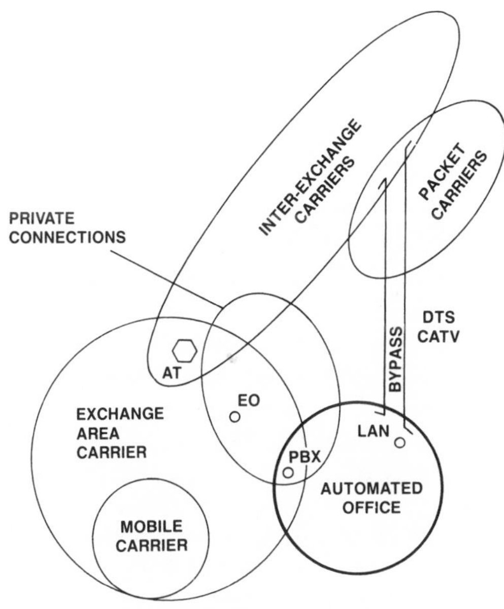

Figure 5.7. External telecommunication connections to an automated office.

5.2.2.1. *Exchange Area Facilities*

Exchange area facilities provide access to

- on-demand, switched voice connections to virtually anywhere in the world,
- on-demand switched data connections (such as CSDC and LADT) to an increasing number of points in the United States and the developed world, and
- arrangeable connections for digital videoconferencing and high-speed data to many points in North America.

They can also provide switched voice connections to mobile users. An increasing number of optical fibers are being placed in the local loop: they will be used to support wideband and high-bit-rate connections for major customers.

5.2.2.2. *Bypass Facilities*

Bypass facilities provide connections from a local distribution point, such as an earth station, to the customer's premises without involving facilities of the local telephone company. Most often, the connections complete an end-to-end digital circuit made through an interexchange carrier. They will be implemented by DTS links (see Section 4.7.6) or CATV channels (see Sections 4.4.5.3 and 4.4.5.4). Short distances may be bridged by free-space laser beams or optical fibers.

5.2.2.3. *Long-Distance Connections*

Long-distance connections are made by interexchange carriers and packet carriers who connect with exchange area carriers at an exchange area tandem or a local node on which bypass facilities terminate. Connections may also be made directly to the office site through an on-premises earth station provided by a special carrier or owned privately. The range of connections and equipment employed in providing telecommunications connections within and outside the automated office is summarized in Figure 5.8.

5.2.3. Security

The conduct of business is based on orders, contracts, and instructions authenticated by signatures or other code marks which certify that responsible persons have created and accepted these transactions. In cases of dispute, the documents can be inspected for authenticity and forgeries and alterations can be detected. The conduct of business also depends on accurate data, correct records, and an ability to shield sensitive information from competitors. Authentication, integrity of information, and confidentiality are essential to every enterprise.

CONNECTION		COMMUNICATIONS REQUIREMENTS FOR AUTOMATED OFFICE				
		VOICE ONLY	DATA ONLY	VOICE AND DATA	VIDEO ONLY	VOICE, DATA AND VIDEO
INTRA—OFFICE	1.	Wire	Wire	Wire	Cable/Fiber	Probably
	2.	Analog/Digital	Digital	Digital	Analog	not
	3.	Star	Bus/Ring	Star/Bus	Star/Tree	combined
	4.	Circuit	Packet	Circuit/Packet	Circuit	in office
	5.	Telephone	Data Terminal	Workstation	Video Display	
INTER OFFICE, INTRA BUILDING	1.	Wire	Wire/Cable	Wire/Cable	Cable/Fiber	Cable/Fiber
	2.	Analog/Digital	Digital	Digital	Analog	Analog and Digital
	3.	Star	Bus/Ring	Star/Bus	Star/Tree	Star
	4.	Circuit	Packet	Circuit/Burst /Packet	Circuit	Star
	5.	PBX	LAN	PBX/LAN		
INTER BUILDING, INTRA CITY	1.	Wire/Fiber /Radio	Wire/Cable /Fiber/Radio	Wire/Cable /Fiber/Radio	Cable/Fiber /Radio	Cable/Fiber /Radio
	2.	Analog/Digital	Digital	Digital	Analog/Digital	Analog and Digital
	3.	Star	Bus/Ring	Star/Bus	Star/Tree	Star
	4.	Circuit	Packet	Circuit/Burst /Packet	Circuit	Circuit
	5.	Digital EO	Packet Switch	Burst Switch		
INTER CITY	1.	← Wire / Cable / Fiber / Radio / Satellite →				
	2.	← Digital →				
	3.	← Star →				
	4.	Circuit	Packet	Burst	Circuit	

Figure 5.8. The range of telecommunication facilities which may be required to support an automated office. Items 1 through 5 in each column identify specific parameters. 1 is the most likely transmission medium, 2 is the probable type of signal, 3 is the probable connection topology, 4 is the likely protocol, and 5 identifies associated equipment.

However, the proliferation of electronic techniques in business, the universal availability of computers, and the existence of pervasive telecommunication networks has created an environment in which managers can no longer be completely confident that these conditions prevail.

To protect electronic messages against forgery, fraud, sabotage, and eavesdropping requires the adaptation of existing cryptographic techniques to the electronic environment. Two basic approaches are in use. One, defined by the Data Encryption Standard (DES) developed by IBM and adopted by NBS (National Bureau of Standards), replaces information and scrambles the order of the information to convert plain text into ciphertext. The operations of substitution and transposition are achieved through the use of a 64-bit word which contains a 56-bit cypher key and eight parity bits to detect transmission errors. The cypher key is secret and is used for both encryption and decryption. To operate a network, the same secret key must be available to all users. Widespread distribution of the key poses a threat to the security of the operation.

A second approach, called the *public key* system may be safer, and is easier to manage. In public key cryptography, encryption and decryption are performed on electronic messages using two keys. One is a *public* key, the other is a *secret* key. Either can be used for encryption: decryption is achieved through the use of the other key. Each user of the system generates a public key (PK) and a related secret key (SK). Public keys are known to other users. Thus, when user 1 wishes to send a private message (M_{12}) to user 2, it is encrypted as $PK_2.M_{12}$. When user 2 wishes to read the message, it is decrypted as $SK_2.PK_2.M_{12}$ to yield M_{12}. Since decryption is the inverse of encryption, knowing PK_2, M_{12}, and $PK_2.M_{12}$ it should be possible to calculate SK_2. In practice, while the keys are generated together from a single key generation algorithm, this algorithm is chosen so that calculating SK and PK independently is hard to do. The degree of difficulty of this calculation is a measure of the degree of security provided by the particular algorithm.

In the example above, it was necessary for the sender (user 1) to use public information and for the recipient (user 2) to employ secret information to obtain privacy. To obtain authentication, the sender (user 1) must employ secret information and the recipient employs public information. Thus, if user 1 wishes to send a message M_{12} to user 2, and wishes user 2 to know without doubt that it comes from user 1, the message is encrypted using user 1's secret key as $SK_1.M_{12}$. When user 2 receives the message, application of user 1's public key produces message M_{12}. To obtain both privacy and authentication, user 1 encrypts the message M_{12} as $SK_1.M_{12}$ and then as $PK_2.SK_1.M_{12}$. On receiving the message, user 2 applies SK_2 to obtain the message $SK_1.M_{12}$ and then applies PK_1 to obtain M_{12}.[6] The principles of DES and public-key cryptosystems are illustrated in Figure 5.9.

5.3. Point-to-Multipoint and Broadcast Facilities

In contrast to the intercommunication environment of digital voice, data, and teleconferencing which is present in the business sector, the mass communication environment of the residence sector does not afford the same opportunity to reduce all signals to a common (digital) format. This is due to the fact that broadcasting is dependent on the availability of electromagnetic spectrum space for clear channels over which to send information to all who wish to receive it. Because the spectrum is finite and competition for frequency allocations shows no sign of diminishing, the current channel widths assigned to radio and television broadcasting are unlikely to be increased. To provide the digital equivalent of the information signals employed in AM radio, FM radio, and TV broadcasting (using the present bandwidths) would require the introduction of digital modulation techniques with spectral efficiencies of many bits/second/hertz. To achieve such values reliably in the multipath environment of the average urban area would be next to impossible. Broadcasting, then, will

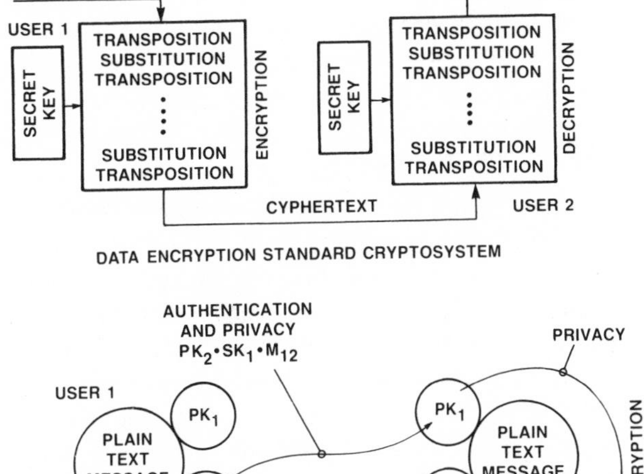

Figure 5.9. Principles of encryption and decryption of data using data encryption standard (DES) and public-key cryptosystems.

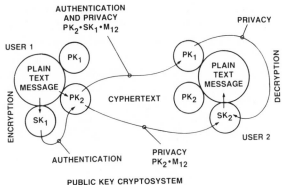

continue to use analog modulation techniques while employing digital technology wherever possible to improve the performance and reduce the cost of studio equipment and receivers (see Section 4.4.2). Because of the need to be compatible with broadcast receivers, it is unlikely that CATV systems will press for digital television signals. However, the other services the systems provide (such as interactive services, point-to-point data transport, information services, etc.) will employ digital technology wherever possible.

5.4. Residence Services

Future residence services will incorporate microcomputer-based entertainment and data processing functions and residential communications media (such as telephone, radio, and television). They will take advantage of the new communications media described in Section 2.3 to provide an environment in which entertainment, information, and personal communication are readily available, and some administrative, security, and conservation functions can be performed automatically. Many of the applications will require development of specialized infrastructures (e.g., networks, databases, administrative centers) which will have a social or commercial basis.

5.4.1. The Wired Household

The concept of an all-encompassing household communications–information system is shown in Figure 5.10. It can be called a *wired* household. It includes arrangements in which security, entertainment, control, and communications functions are performed. The equipment includes a broad spectrum of household devices; information and entertainment, command and control and administration subsystems based on a household computer; and local communications facilities. This concept makes use of telecommunication connections which may be supplied by the local telephone carrier over wires or fibers, the local CATV company over cables or fibers, and (perhaps) the local electric utility over existing power wiring, as well as antennas which receive transmissions from local radio and television broadcasting stations, and television signals broadcast directly from space by satellite. Communication within the household may

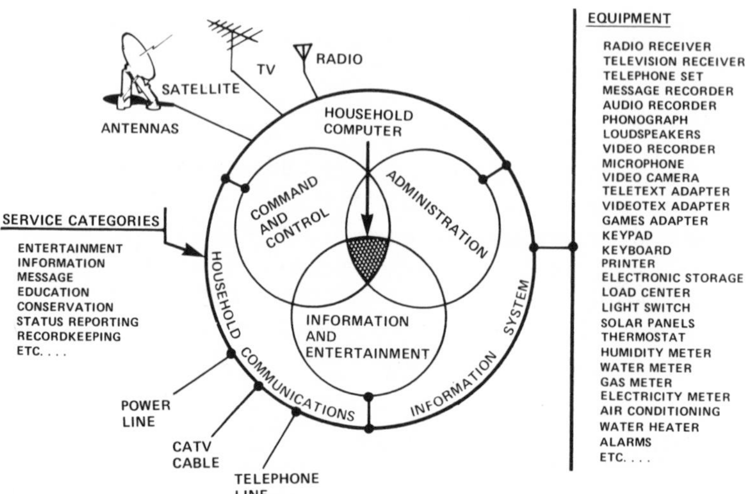

Figure 5.10. The elements of a wired household, a concept that includes a home computer, enhanced communications, and new information sources to provide an environment in which entertainment, information, and personal communications are readily available, and administrative, security, and conservation functions can be performed automatically. Teletext, Viewdata, and Interactive CATV have applications in this environment. © 1981 IEEE.

be completed over wire or cable, by coded messages on the power wiring or wireless links using sonic, electromagnetic, or infrared energy. Besides the familiar uses of the radio, television, and telephone, some of the functions which each subsystem could perform are as follows:

Information and Entertainment. Provide retrieval of catalogs, schedules, and library information, news and reports, using equipment such as a television receiver, videotex decoder, telephone, modem, and a keypad or keyboard/printer. These would be supported by broadcast videotex services, wired videotex services, community services, or other information sources.

Provide interactive education for preschool, in-school, vocational and continuing students, and interested adults, using equipment such as a television receiver, video and/or audio recorder, microphone, television camera, and a keypad or keyboard/printer. These would be supported by prerecorded material or community services.

Provide interactive games and intellectual entertainment for children and adults, using equipment such as a television receiver, television games attachment, video and/or audio recorder, a keypad or keyboard/printer. These would be supported by prerecorded material, community services, or other subscribers.

Provide interactive entertainment, opinion polling, preference sampling, and voting, using equipment such as television receiver and keypad, supported by broadcast and cable services.

Command and Control. Adjust electrical load by time-of-day remote command from utility. Provide meter information to utility on-demand, or at preset intervals. Optimize use of solar panels, air conditioning, and space heaters, to maintain living environment within preset temperature and humidity limits, yet conserve energy consumption. Monitor fire, intrusion, and assistance alarms, notifying emergency services or community center as appropriate. Provide system status information to remote caller (householder). Turn on lights, radio, and heat on command (local or remote) or in accordance with a preset scenario.

Administration. Provide interactive information retrieval using equipment such as television receiver, video, and/or audio-recorder, household data base, a keypad or keyboard/printer. These would be supported by prerecorded material, remote data centers, or community services. Maintain family records such as accounts, medical history, addresses, and telephone directory. Pay bills by electronic funds transfers. Compute taxes. Send messages to and receive messages from others (electronic mail).

Other items of household equipment (such as washer, dryer, oven, freezer, water and gas meters,) could be included and more specialized features incorporated. Perhaps it is sufficient to note that a computer and associated equipment can automate almost all household functions requiring intellectual (as opposed to physical) activity, and they can pass beyond the individual household when supported by appropriate local data banks and community services, or even

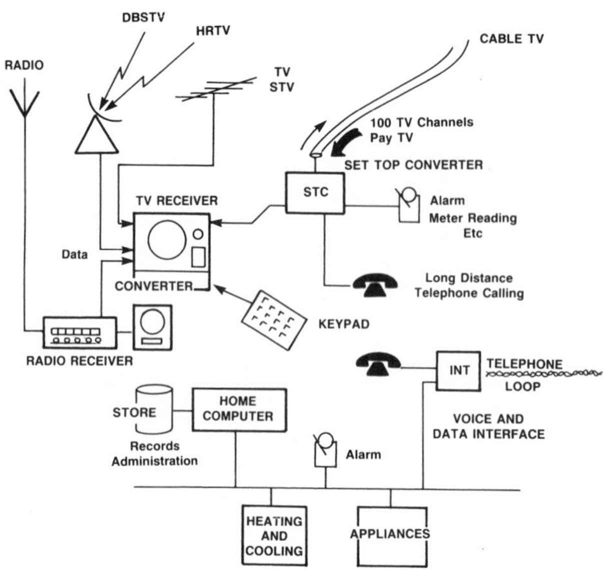

Figure 5.11. Some of the information, entertainment, security, and administration options that may be available in the wired household.

nationwide services (perhaps distributed by satellite). To implement the full package with today's technology would be enormously expensive, but possible. Future technology will make the task easier, and if every household were to be so equipped, economies gained from volume production might make the cost affordable.

The ideas embodied in Figure 5.11 illustrate the potential of advanced technologies to provide additional communications–information services which might be attractive to consumers and providers. Such unifying models can be helpful to the working technologist, but it would be a mistake to attribute a reality beyond this and assume that a total system *must* emerge. To do so would be to ignore the influence of existing interests and the importance of existing facilities. The idea of one integrated, universal system flies in the face of the present competitive fragmentation of providers in the United States. Many of them see opportunities to extend their present markets and exploit the base provided by their existing products.

Parts of this system will be developed and tested by those having the most to gain. Thus, the consumer electronics industry will exploit the television receiver and provide adapters which make it a color display for games and information. Cable television companies will take advantage of their wideband connection to supply additional services, including return services. Those who supply home computers may emphasize command and control functions, information retrieval, and data storage. Local telephone companies, and others, may offer electronic directories and catalogs as well as remote control and access features. Certainly, one-way video signals for entertainment and two-way voice signals over the telephone network for personal messages will be universally required by customers. However, the emphasis placed on data connections for information retrieval, remote control and monitoring, meter reading, and other services will depend on the character of the residential activity. A family with growing children may emphasize education; professional adults may emphasize information, record keeping, etc.; and a household of many busy persons may emphasize message and reporting services. Most customers may be receptive to conservation-supporting services.

In the wired household, users will have an assortment of terminals available to them which respond to two-way voice, two-way data, one-way audio, and one-way video messages. Direct connections will probably be made to telephones and television receivers. Functions such as climate control, meter reading, security alarms, and lighting controls may be supervised by a central system such as a home computer. Wire or wireless (sonic, radio, and infrared radiation) connections will be used for these connections. Personal computers which provide the professional with computing capability extendable to household record keeping, information systems, and financial management, may be connected through external bus arrangements to outside networks making energy management, message handling, and other functions available. The development of high-level, user-friendly, programming tools and packages will place these capabilities in the hands of virtually everyone who wishes to have them.

It remains to be seen whether the majority of households will have a need for centralized control, management, and communications systems—or whether specialized units performing individual tasks will be more popular. Figure 5.11 suggests some of the options which might be available. At least one serious proposal has been made for a low-cost network for the exchange of control information among a broad spectrum of consumer applications in the home.[(7)]

5.4.2. Facility Integration

In the mixed environment surrounding the wired household, integration of the facilities is more difficult than in the business services environment of the automated office. Nevertheless, some sharing is possible among those services which are delivered by wire or cable. The question facing the supplier of mixed services is how to provide many one-way wideband downstream channels (both point-to-multipoint and point-to-point) in combination with one or two two-way voice channels, and to do it cost effectively. While this is a difficult problem, it would be much worse if return video was also required. Fortunately, the cost of video equipment can be relied on to be a strong deterrent to any large-scale origination of video in the household (including video telephone) for some time to come.

The degree of integration which is appropriate in a specific situation depends on many factors, not the least of which is the state of existing facilities. For instance, when telephone and CATV already exist, a straightforward approach to providing limited mixed services is to use the telephone as a return channel in response to downstream television activity. Physical integration may involve the sharing of local transmission facilities, so that all signals enter the household by the same route. The integration of local transmission media implies more than a common electrical/optical pathway. It includes multiplexing and demultiplexing equipment, as well as arrangements for sharing some terminal capabilities. The number of video channels devoted to entertainment and general information (one-to-many channels) affects the transmission media which can be employed. For short distances (a few hundred meters), twisted pair may be used for a limited amount of video and voice. For longer distances, coaxial cable or glass fibers must be employed. The topology is determined by the need to provide two-way point-to-point channels for voice and data and the ability to record and bill for special services (such as pay-TV, teleshopping, etc.) These features can be obtained through a switched optical fiber distribution system which delivers entertainment and information video as well as two-way voice and data channels. However, the distance optical fibers carry analog video signals is something less than the distance present wire pairs carry voice so that the existing telephone network layout cannot be used as a model. Rather, additional distribution centers (local distribution centers, LDCs) must be introduced where point-to-point and point-to-multipoint circuits can be combined into the final transmission link to the household. To reach the LDC, entertainment and information channels can be connected from the CATV head-end over fiber or cable trunks, and the voice and data channels can be connected to the local telephone end office or to bypass facilities.

5.4.3. Integrated Services Field Trials

Several administrations have installed small field trials of integrated services systems which provide a mixed voice, data, and video environment, and employ optical fibers. In the main, they are intended for household use. The principal objectives of these trial activities are as follows:

- to gain experience in the design and implementation of optical fiber based customer access networks;
- to provide a test bed for new telecommunication services to determine the information needs of customers; and
- to develop new product and service concepts.

In Elie, Manitoba, the Department of Communication of the Government of Canada, the Canadian Telecommunications Carriers Association, the Manitoba Telephone System, Infomart, and Northern Telecom sponsored an 18-month trial of an integrated services fiber optics distribution system serving 150 households. Telephone, television, stereo FM radio, and data were provided over a switched star network. A feature of the trial was the availability of Telidon services supported by databases providing agricultural (18,000 pages), lifestyle (12,000 pages), and government (90,000 pages) information, as well as computer-aided instruction and reference libraries. The three most used Telidon services were electronic games, weather and news, and messaging. The most frequent users were children, followed by adult males and adult females. The 18-month trial was completed in 1983. It has been decided to continue its operation as a national test bed for future trials of new services.[(8)]

In 1983, in ten locations in the Federal Republic of Germany, the Deutsche Bundespost installed optical fiber local loops using system concepts developed by six different suppliers (or groups of suppliers). Called BIGFON, the project incorporates all existing narrowband services in the public network, as well as videotelephone, television, and stereo radio channels. In all, some 350 subscribers are involved in six cities. Eventually, BIGFON islands will be linked by long-haul networks.[(9)]

In Japan, a fiber optic broadband system has been in operation in Higashi-Ikoma New Town since 1977. Called Hi-OVIS (highly interactive optical visual information system), it provides advanced CATV and interactive television services over a star connected fiber network to 158 household terminals and 10 terminals in public institutions. The test seeks to meet certain social requirements as well as to test optical fiber distribution techniques. A second field trial—concentrating on optical fiber customer loops—was conducted in Yokosuka in 1980–1981. It demonstrated the feasibility of business loops distributing voice, video conferencing, and data services, and of household loops distributing voice, advanced CATV, and data services. The loops employed wavelength division multiplexing (WDM) to achieve two-way transmission over a single fiber to each customer. A third system, demonstrated in Kobe in 1980, provided videotelephone and interactive visual information services over single fibers employing WDM.[(10)]

In France, at Biarritz, the DGT (Direction Générale de Télécommunications) has begun a major trial with 1500 customers, some located in a densely populated section in the center of town and others in a residential district with low population density. The facilities being installed are said to have an ultimate capacity of 5000 customers. The system offers switched and distributed services including voice telephone, videotelephone, television channels, stereo radio, and videotex. This, the largest broadband fiber optic trial to date, is intended to provide the French telecommunications industry with experience in the design, installation, and operation of fiber optic systems, to evaluate customer acceptance of advanced services, and to stimulate commercial production of fiber optics components and equipment.[(11)] The DGT is experimenting with a videotex-based electronic directory service in the town of Velizy.

5.5. The Future

Written and updated over a period of two years, this book describes telecommunication media and facilities as they have been shaped by the technology of the middle years of the 20th century. It is a snapshot of human achievement which contains a finite amount of information for reference and use in building the facilities and media of tomorrow (see Section 2.1.1.1). Before discussing the future thrusts of telecommunication in the 801st lifetime (see Section 1.2.1), we shall outline the near-term developments in the United States, and in the rest of the world.

5.5.1. United States

A combination of technology push and market pull (see Section 1.3.1) will shape the services available and determine the structure of an industry which is changing, and must continue to change, as the needs of its customers develop and the rewards available to the providers become apparent. Who will design and manufacture the equipment, operate and maintain the facilities, develop and implement the services, and use them for telecommunication depends upon the future course of technology and the economics of competition.

5.5.1.1. Major Participants

As discussed in Chapter 1, telecommunication services in the United States used to be provided by corporate entities operating in well-defined service areas. Through the Bell System, AT&T supplied point-to-point voice services to the majority of the American public. Three television networks, CBS, ABC, and NBC, supplied one-to-many, commercially sponsored television programs through chains of local broadcasting stations, and point-to-multipoint cable systems enabled households situated in remote regions, or in difficult terrain, to receive these signals. A myriad of broadcasters supplied one-to-many commercially sponsored radio programs, and a small number of *public* broadcasters supply noncommercial radio and television. Now, the technology engine started by World War II has created a growing need for data communications, presented an irresistible economic case for the use of digital techniques, made opportunities for enhanced services, and forced competition into the telecommunication marketplace. As a result, telecommunication services are supplied by a host of companies. Some of the major organizations providing telecommunication services in the local area are listed in Figure 5.12.

By far the largest organizations are those companies which provide exchange area services. The seven regional operating companies formed from the 22 operating companies of the Bell System, GTE Corporation, United Telecom, Continental Telecom, and Southern New England Telephone serve approximately

EXCHANGE AREA SERVICE COMPANIES		Approximate % US Customers (1984)
Nynex		12
Bell Atlantic		13
Bell South		12
Ameritech		13
Southwestern Bell		9
US West		10
Pacific Telesis		10
GTE		9
United Telecom		3
Continental Telecom		2
SNET		1
		94
MOBILE SERVICE COMPANIES		
Exchange Area Service Companies		
GTE Mobilnet		
Metromedia		
MCI		
BROADCASTERS		
CBS		
ABC	and	
NBC	Affiliated	
Metromedia	Companies	
PBS		
CATV COMPANIES		
TCI	Cox Cable	
ATC	Storer Broadcasting	
Group W Cable	Warner-Amex	
DIRECT BROADCAST SATELLITE TV COMPANIES		
United Satellite Communications		
Comsat		
RCA		
CBS		

Figure 5.12. Listing of some major organizations providing telecommunication services in local areas.

95% of U.S. telephone customers. Less than 10% of these customers account for two-thirds or more of all revenues derived from basic telephone service. They are the prime targets for bypass by CATV and DTS carriers. The television networks are in competition with pay-TV over cable, broadcast subscription-TV, and with direct broadcast satellite companies.

Long-distance telecommunication services are provided by several companies. The single dominant corporation is AT&T Communications Corporation, which operates interexchange switching and transmission facilities (ATTIX) previously owned by the BOCs and the Long Lines Department of AT&T. Alternative long-distance services are offered by MCI, GTE Sprint, and others. A list of some of the organizations participating in long-distance telecommunication is given in Figure 5.13. There is vigorous competition among satellite service companies and packet network companies for voice and data traffic from the business sector.

INTEREXCHANGE SERVICE COMPANIES	Approximate % of US Market (1984)
ATT Communications	93
MCI	4
GTE Sprint	2
	99

SATELLITE SERVICE COMPANIES

- ATT Communications
- GTE Spacenet
- RCA
- SBS
- Western Union

PACKET NETWORK COMPANIES

- ATT Information Systems
- Tymnet
- GTE Telenet

Figure 5.13. List of major organizations providing long-distance telecommunication services.

5.5.1.2. The Freedom to Compete

In an environment in which technical advances are made every day and each new discovery leads to more than one fresh idea, conditions are ripe for an *explosion*—a technology explosion which creates a vast array of technical ideas seeking markets. The freedom to compete for even a small piece of a very large and growing telecommunication market ensures that entrepreneurial enterprise will exploit whatever developments have a chance of providing better performance for the customer and of returning a profit to the provider. For established companies with a large investment in facilities, the constant challenge is to reprice and upgrade existing services, so as to return the competition if it threatens the stability of existing revenue sources or if it turns out to satisfy a permanent need (thereby seizing an opportunity to expand revenue sources). In the long run there are limits: repricing must be based on costs incurred, and the extent to which it is possible to upgrade depends on the basic techniques incorporated in the in-place equipment. Digital facilities can be changed out piece-by-piece to increase storage capacity, bus speed, and processing power, and software can be modified to incorporate new features, particularly if the source employs a high-level language and is structured. With analog facilities the same fundamental flexibility is not present, although modular design and linear IC technology provide some help.

Today, all major cities have access to two or more interexchange carriers. All of them are expanding capacity and extending their geographic coverage, adding satellites, terrestrial optical fibers, and other facilities as rapidly as capital can be raised and equipment can be designed and manufactured. Dedicated digital channels at 56 kb/s and 1.54 Mb/s are available to those with the need, and circuit-switched digital capability (up to 3 Mb/s) is spreading.

CATV companies located in commercial centers are constructing dual systems: one to service their traditional residential entertainment customers, and the other to supply the data needs of business. A 300 MHz CATV system can serve the data transport needs of over 100,000 medium- to low-speed terminals and, by 1990, transport facilities (many of them fiber) owned by CATV companies are expected to pass by at least half of all major data processing centers in the 50 largest SMSAs. Data transport over optical fibers is available in some of the financial and commercial districts of large cities connecting to *teleports* where earth stations direct messages to a band of satellites from which they can be radiated to the rest of the country and much of the world.

Within a year or two, almost 40 communications satellites will be in geostationary orbit across the United States. They will be used for many purposes associated with commerce, including nationwide services such as videoconferencing, video seminars, paging, and similar activities. They will also be used for the direct broadcasting of standard television and, perhaps, a high-definition service. Their signals will compete directly with those provided by terrestrial broadcasters and CATV companies. Taking advantage of satellite transmission from program sources to head end, and switched, optical fiber distribution, CATV systems can now offer well over 100 channels. This variety of local, regional, and national programs cannot be matched by either terrestrial or space broadcasters.

The freedom to compete assures American business will be able to obtain the telecommunication support essential to the growth of office automation. It may be that only the United States has an economy of sufficient breadth, vigor, and geographic extent to support the proliferation of these services and the organizations to provide them. The supply of facilities is another matter. The freedom to compete assures that the best equipment will be procured from wherever in the world it is produced. Many foreign companies have the capital to invest in leading edge development projects, and are anxious to sell their products in the largest telecommunication market in the world. As far as household communications are concerned, developments depend on whether the information-oriented business environment penetrates the average home and creates a demand sufficient to support the cost of special terminals which can make use of the common facilities already installed for business purposes in nonbusiness hours. Notwithstanding the demonstration projects undertaken by many telephone administrations, only those household services which can be profitable will proliferate.

5.5.2. The Rest of the World

In the rest of the world, the same degree of freedom to compete in the provision of telecommunication services does not exist. The exceptions could be those developed countries which aspire to a free enterprise economy to some significant degree. Developed through competition in the United States, advanced business services which have a permanent value will probably be introduced by the cognizant agencies in response to demands to modernize industry and commerce. Even though each country may elect to have its own variations, great emphasis will be placed on CCITT Recommendations. This organization may be unable to keep up with the pace of technical developments.

In Canada, Japan, Great Britain, France, and West Germany, some households are testing the delivery of telecommunication services over optical fibers, and in these and other countries videotex services (both teletext and viewdata) are in use. Direct broadcast satellites are in the planning stage in Europe. At present, these activities appear to be sustained largely by a continuing technology push from the technical staffs of the telecommunication agencies.

5.5.3. Telecommunication in the 801st Lifetime

The 801st lifetime stretches to the middle of the 21st century. A child born in the United States today will grow to adulthood taking pervasive telecommunication for granted and will regard as new only the entries which have yet to be written on the right-hand side of Figure 1.2 (see Chapter 1). Bearing in mind the changes from the 1920s to the present (the 800th lifetime), it would be foolhardy to attempt to fill them in. Exploding technology will provide all manner of inventions which will lead to innovations based on microstructures, software, artificial intelligence, and other new technical fields. Electronics and optics will dominate physical implementation, and artificial intelligence will be the driving force behind customization through the application of software. Direct speech input facilities and expert systems will evolve to make the person and machine interface easier to use and to ensure constructive dialog between person and machine. Truly personal telecommunication will be possible through

- pocket-size terminals, or terminals which attach to the wrist or ear for telephone, television, information, mail, and other services;
- broadcast electromagnetic media which make the services available virtually anywhere; and
- customized offerings which satisfy individual needs for communication, information, and entertainment.

Truly high-bit-rate (wideband) telecommunication will be possible through the use of optical technology. Streams with a rate of 1 Gb/s, which can transport the text of the *Encyclopedia Brittanica* in approximately ten seconds, will be freely available. At the option of the user, real-time, or store-and-forward capability will be able to be invoked so that reliable voice, data, or video telecommunication can be achieved irrespective of schedules and time zones. These capabilities will make new versions of old activities possible, such as

- teleshopping: shopping from home, or another location during normal hours, or at any hour;
- telebanking: banking, bill-paying, etc. from a convenient location at any hour;
- telehousekeeping: controlling or monitoring household functions from another location;
- telemedicine: supervising or monitoring medical situations from another location;
- tele-education: personal study and instruction at home or another location; and
- telework: performance of work tasks at home or another location.

These arrangements contribute to user convenience and can increase the productivity of the provider. However, because the personal interaction of user and provider is eliminated, they may not suit every situation. For instance, shopping from home can be about the same as browsing through a catalog—but how much better it is to see and feel the merchandise—and shopkeepers are dependent on the persuasive powers of their salespersons and the sudden impulse of the customer to buy something not thought of before entering the store. (As discussed in Section 2.3.3.7.2, it is much easier to say *no* as the richness of the medium decreases.) Automated teller machines (ATMs) dispense money at the user's convenience, but retail banking is a pursuit founded on depositors' goodwill—and one ATM is much the same as another. Good medicine requires personal contact. The right teacher makes education much more than learning facts. Working at home deprives the participant of the stimulation of co-workers. Nevertheless, some versions of these arrangements will be implemented.

If the trends of the last two decades continue, there will be an increasing number of households which are unoccupied during the day (due to an increasing number of families in which both spouses work, and a decreasing number of children). Such units might make use of shopping, banking, and housekeeping services. There will also be an increasing number of households occupied by elderly persons. They might make use of shopping, banking, and medical services. For those households which are occupied during the day by mothers with children, education and work services might be attractive. For all it is likely that telecommunication will assume a greater role as social surrogate through the provision of entertainment and of links to the outside world.

Telecommunication, per se, is not a primary force in the world. It shapes neither political theory nor political boundaries nor political reality. Rather, telecommunication facilitates all three by contributing to the infrastructure which supports the modern world—and it is affected by them. In the United States, a procompetitive, deregulation thrust is well established and is being implemented. The future results and implications are already evident to political theorists. They assert that the burgeoning telecommunication capabilities will provide an important and essential element of the infrastructure of the information society. Intimately entwined with computers, telecommunication will contribute to the growth of a culture dependent on timely access to information drawn from many sources and organized according to individual needs and to the growth of an information culture which is literate and aware of events around the world. For political realists, the effects have yet to be measured and evaluated. Whatever happens, it is likely that new dimensions will continue to be added to the telecommunication experience guided by customer needs and political opportunities. One thing is certain, there will be more than enough technology to implement whatever is required!

References

Chapter 1

1. Toffler, A., *Future Shock,* Random House, New York (1970).
2. Utterback, J.M., Innovation in industry and the diffusion of technology, *Science* **183,** 620–626 (1974).
3. Ellis, L.W., Scale economy coefficients for telecommunications, *IEEE Trans. Syst. Man Cybern.* **SMC-10**(1), 8–16 (1980).
4. Mason, L.G., and Combot, J.-P., Optimal modernization policies for telecommunications facilities, *IEEE Trans. Commun.* **COM-28**(3), 317–324 (1980).
5. United States vs. Western Electric Co., Inc. and American Telephone and Telegraph Co. 1982–2 Trade Cases (CCH)§64.900.

Chapter 2

1. McLuhan, M., *Understanding Media: The Extensions of Man,* McGraw-Hill, New York (1964).
2. O'Brien, C.D., and Brown, H.G., A perspective on the development of videotex in North America, *IEEE J. Sel. Areas Commun.* **SAC-1**(2), 260–266 (1983).
3. Teramura, H., Ono, K., Ando, S., Yamakazi, Y., Yamamoto, S., and Matsuo, K., Experimental facsimile, communication system on packet-switched data network, *IEEE Trans. Commun.* **COM-29**(12), 1942–1951 (1981).
4. Johansen, R., McNeal, B., and Nyhan, M., Telecommunications and developmentally disabled people, Institute for the Future, Menlo Park, California, Report R-50 (1981).
5. Chapanis, A., Interactive communication: A few research answers for a technology explosion, *8th Annual Conference of the American Psychology Association, Toronto, Ontario, Canada,* Johns Hopkins University, Baltimore, Maryland (1978).
6. Hansell, K.J., Green, D.L., and Erbring, L., Survey examines benefits of videoconferencing, *Telephony* **203**(4), 37–39; (1982) and Videoconferencing boosts productivity, users report, *Telephony* **203**(5), 38–40, 44, 69 (1982).
7. *On the Cable: The Television of Abundance,* Report of the Sloan Commission on Cable Communications, McGraw-Hill, New York (1971).

Chapter 3

1. Keshner, M.S., 1/*f* noise, *Proc. IEEE,* **70**(3), 212–218 (1982).
2. Oppenheim, A.V., and Lim, J.S., The importance of phase in signals, *Proc. IEEE,* **69**(5), 529–541 (1981).
3. Seidl, R.A., A tutorial paper on medium bit rate speech coding techniques, *Aust. Telecom. Res.* **17**(1), 61–72 (1983).
4. Oetting, J.D., A comparison of modulation techniques for digital radio, *IEEE Trans. Commun.* **COM-17,**(10), 1752–1760 (1979).
5. Pickholtz, R.L., Schilling, D.L., and Milstein, L.S., Theory of spread-spectrum communications—A tutorial, *IEEE Trans. Commun.* **COM-30**(5), 855–884 (1982).
6. Rothmaier, K., and Scheller, R., Design of economic PCM arrays with a prescribed grade of service, *IEEE Trans. Commun.* **COM-29**(7), 925–935 (1981).
7. Nesenbergs, M., A hybrid of Erlang B and C formulas and its applications, *IEEE Trans. Commun.* **COM-27**(1), 59–68 (1979).
8. Enos, J.C., and van Tilburg, R.L., Software design, *(IEEE) Computer* **14**(2), 61–84 (1981).
9. Special Issue on Fifth Generation Computing, *IEEE Spectrum,* **20**(11), (1983).
10. Rabiner, L.R., Note on some factors affecting performance of dynamic time warping algorithms for isolated word recognition, *Bell Syst. Tech. J.* **61**(3), 363–373 (1982).
11. Wilpon, J.G., and Rabiner, L.R., On the recognition of isolated digits from a large telephone customer population, *Bell Syst. Tech. J.* **62**(7), 1977–2000 (1983).
12. Rabiner, L.R., and Levinson, S.E., Isolated and connected word recognition—Theory and selected applications, *IEEE Trans. Commun.* **COM-29**(5), 621–659 (1981).
13. O'Shaugnessy, D., Automatic speech synthesis, *IEEE Commun. Mag.* **21**(9), 26–34 (1983).
14. Long, S.I., Welch, B.M., Zucca, R., Asbeck, P.M., Lee, C.-P., Kirkpatrick, C.G., Lee, F.S., Kaelin, G.R., and Eden, R.C., High-speed GaAs integrated circuits, *Proc. IEEE,* **70**(1), 35–45 (1982).
15. Nuzillat, G., Perea, E.H., Bert, G., Damay-Kawala, F., Gloanec, M., Peltier, M., Ngu, T.P., and Arnodo, C. GaAs MESFET ICs for gigabit logic application, *IEEE J. Solid-State Circuits,* **SC-17**(3), 569–583 (1982).
16. Eden, R.C., Livingston, A.R., and Welch, B.M., Integrated circuits: The case for gallium-arsenide, *IEEE Spectrum,* **20**(12), 30–37 (1983).
17. Arnold, J., FET technology for low-noise front ends, *IEEE Commun. Mag.* **21**(6), 37–42 (1983).
18. Mitachi, S., Ohishi, Y., and Miyashita, T., A fluoride glass optical fiber operating in the mid-infrared wavelength range, *(IEEE) J. Lightwave Technol.* **LT-1**(1), 67–70 (1983).

Chapter 4

1. Special Issue on AR6A Microwave Radio, *Bell System Tech. J.* **62**(10), Part 3 (1983).
2. Jacobs, I., and Stauffer, J.R., FT3—A metropolitan trunk lightwave system, *Proc. IEEE,* **68**(10), 1286–1290 (1980).
3. Reudink, D.O., Acompora, A.S., and Yeh, Y.-S., The transmission capacity of multibeam communications satellites, *Proc. IEEE* **69**(2), 209–225 (1981).
4. Emling, J.W., and Mitchell, D., The effects of time delay and echoes on telephone conversations, *Bell Syst. Tech. J.,* **42**(9), 2869–2891 (1963).
5. Ribeyre, C., Prat, M., Maitre, X., and Coullare, P., Exploitation of transmultiplexers in telecommunication networks, *IEEE Trans. Commun.* **COM-30**(7), 1493–1497 (1982).
6. Special Issue on No. 4 Electronic Switching System, *Bell Syst. Tech. J.* **60**(6) (1981).
7. London, H.S., and Guiffrida, T.S., High-speed switched digital service, *IEEE Commun. Mag.* **21**(2), 25–29 (1983).

8. Giesielka, A.J., Greco, G.J., Hawley, G.T., Sevdinoglon, A.C., and Stevens, L.J., The GREG system plan, *IEEE Trans. Commun.* **COM-28**(7), 931–942 (1980).
9. Cooper, G.R., and Nettleton, R.W., A spread-spectrum technique for high-capacity mobile communication, *IEEE Trans. Veh. Technol.* **VT-27**(11), 264–275 (1978).
10. Cooper, G.R., and Nettleton, R.W., Cellular mobile technology: The great multiplier, *IEEE Spectrum,* **20**(6), 30–37 (1983).
11. *Future Land Mobile Telecommunications Requirements,* Final Report, Private Radio Bureau, Federal Communications Commission, Washington, D.C., pp. 5-19–5-26 (1983).
12. *Future Land Mobile Telecommunications Requirements,* Final Report, Private Radio Bureau, Federal Communications Commission, Washington, D.C., pp. 5-26–5-30 (1983).
13. Reudink, D.O., Estimates of path loss and radiated power for UHF mobile-satellite systems, *Bell Syst. Tech. J.,* **62**(8), 2493–2512 (1983).
14. Ikoma, M., Orikasa, H., Suzuki, N., Yano, K., Irema, T., and Shimasaki, N., Digital rural subscriber system, *4th World Telecommunication Forum, Part 2,* International Telecommunication Union, Geneva, **2** 2.5.5.1–2.5.5.8 (1983).
15. Davis, J.R., Forman, L.E., and Nguyen, L.T., The metropolitan digital trunk plant, *Bell Syst. Tech. J.* **60**(6), 933–964 (1981).
16. Puccini, S.E., and Wolff, R.W., Architecture of the GTD-5EAX family, *GTE Autom. Electr. Tech. J.* **19**(3), 110–116 (1980).
17. Abraham, L.G., and Fellows, D.M., A digital telephone with extensions, *IEEE Trans. Commun.* **COM-29**(11), 1602–1608 (1981).
18. Wright, B.A., The design of picturephone meeting service (PMS) conference centers for video teleconferencing, *IEEE Commun. Mag.* **21**(2), 30–36 (1983).
19. Schepers, C., A digital CATV chassis concept, *IEEE Trans. Consumer Electron.* **CE-29**(4), 462–468 (1983).
20. Wetherington, J.D., The story of PLP, *IEEE J. Sel. Areas Commun.* **SAC-1**(2), 167–177 (1983).
21. Rzeszewski, T.S., A compatible high-definition television system. *Bell Syst. Tech. J.* **62**(7), 2091–2111 (1983).
22. Switzer, I., Extended bandwidth cable communications systems, *IEEE Trans. Cable Television,* **CATV-5**(1), 10–17 (1980).
23. Andersen, H.R., and Crane, R.C., A technique for digital information broadcasting using SCA channels, *IEEE Trans. Broadcasting,* **BC-27**(4), 65–70 (1981).
24. Bayer, D.L., and Thompson, R.A., An experimental teleterminal—The software strategy, *Bell Sys. Tech. J.* **62**(2), Part 1, 121–144 (1983).
25. Hagelbarger, D.W., Anderson, R.V., and Kubik, P.S., Experimental teleterminals—Hardware, *Bell Syst. Tech. J.* **61**(1), Part 1, 145–152 (1983).
26. Gfeller, F.R., and Bapst, U., Wireless in-house data communication via diffuse infrared radiation, *Proc. IEEE,* **67**(11), 1471–1486 (1979).
27. Steel, T.B., Jr., The CCITT reference model, *Aust. Telecom. Res.* **16**(3), 15–20 (1982).
28. Hinden, R., Havarty, J., and Sheltzer, A., The DARPA Internet: Interconnecting heterogeneous computer networks with gateways, *(IEEE) Computer,* **16**(9), 38–48 (1983).
29. Stuck, B.W., Calculating the maximum mean data rate in local area networks, *(IEEE) Computer,* **16**(5), 72–76 (1983).

Chapter 5

1. Dorros, I., ISDN, *IEEE Commun. Mag.* **19**(3), 16–19, (1981).
2. Irmer, T., International standards for the ISDN—A challenge for CCITT, *4th World Telecommunication Forum, Part 2,* International Telecommunication Union, Geneva **2,** 2.8.1.1–2.8.1.5. (1983).

3. Musser, J.M., Liu, T.T., and Tredeau, F.P., Packet voice performance on a CSMA/CD local area network, *Proceedings of the International Symposium on Subscriber Loop Systems 82*, pp. 40–44, IEEE, New York (1982).
4. Baxter, L.A., and Baugh, C.R., A comparison of architectural alternatives for local voice/data communications, *IEEE Commun. Mag.* **20**(1), 44–51 (1982).
5. Haselton, F., A PCM frame switching concept leading to burst switching network architecture, *IEEE Commun. Mag.* **21**(6), 13–19 (1983).
6. Muller-Schloer, C., A microprocessor-based cryptoprocessor, *IEEE Micro.*, **3**(5), 5–15 (1983).
7. Gutzwiller, F.W., Francis, J.E., Howell, E.K., and Kruesi, W.R., Homenet: A control network for consumer applications, *IEEE Trans. Consumer Electron.* **CE-29**(3), 297–304 (1983).
8. Akgun, M.B., Harris, K.B., Sigurdson, L., and Tough, G.A., A fibre optic integrated services field trial in Canada, *4th World Telecommunication Forum, Part 2,* International Telecommunication Union, Geneva, **2,** 2.5.1.1–2.5.1.6. (1983).
9. Schenkel K.D., and Braun, E., BIGFON—An all-service subscriber communication system goes into operation, *4th World Telecommunications Forum, Part 2,* International Telecommunication Union, Geneva, **3,** 3.11.7.1–3.11.7.6. (1983).
10. Sakurai, K., and Asataui, K., A review of broad-band fiber system activity in Japan, *IEEE J. Sel. Areas Commun.* **SAC-1**(3), 428–435 (1983).
11. Veilex, R., Experience and projects with fiber optics: An approach for integrating wideband services with local telephone networks, *Supplement to Proceedings of the International Symposium on Subscriber Loop Systems 82,* IEEE, New York, pp. 1–7 (1982).

Glossary

Absorption: attenuation in electromagnetic energy caused by passage through atmosphere, rain, organic materials, and so forth.

Access: point at which entry is gained to a circuit or facility.

Access charge: fee paid to local telephone company by individual customer for opportunity to connect through exchange area facilities to facilities provided by competing interexchange carriers, or by interexchange carrier to local telephone company for opportunity to connect through exchange area facilities to individual customers.

Access tandem: gateway switch connecting exchange area to interexchange facilities.

A/D: analog-to-digital conversion.

Adaptive DM: delta modulation in which step-size adaptation is employed.

Adaptive DPCM: differential pulse code modulation in which the quantizer levels are varied to improve dynamic range and signal-to-noise ratio.

Adaptive PCM: coding technique in which step size is varied to accommodate dynamic range peaks.

Adaptive transform coding: technique that breaks incoming signal into short sequential sections (frames) which are encoded separately on the basis of frequency analysis. A maximum number of bits is assigned to each frame and distributed among the frequency components in accordance with activity.

ADM: adaptive delta modulation; a speech coding technique.

ADPCM: adaptive differential pulse code modulation; a speech coding technique.

A-law companding: a relationship used to shape compression and expansion characteristics of equipment to accommodate the different levels produced by loud and soft talkers. It employs a modified logarithmic relationship for loud and moderate talkers, and a linear relationship for soft talkers.

ALOHA: a pioneering packet system developed at University of Hawaii; also, the protocol employed by this system.

Alternate/alternative route: a second, or subsequent choice path between two switches.

AM: amplitude modulation.

American Wire Gauge: gauge number defines wire of specific diameter, thus, 26 AWG has a diameter of 0.01594 in. (0.0004049 m) and 19 AWG has a diameter of 0.03589 in. (0.0009116 m).

AMI: alternate mark-inversion coding; a bipolar line code.

Amplitude modulation: the amplitude of the carrier wave is varied (modulated) by the amplitude of the signal wave.

Analog signal: a well-defined and continuous signal which may assume any positive or negative values. At all times the value *now* is related to the value that *was* and the value *to be,* by a smooth, continuous variation.

Angle of incidence: the acute angle between a ray and the normal to the surface on which the ray is incident.

APCM: adaptive pulse code modulation.

APD: avalanche photodiode; an optical detector.

ASCII: American Standard Code for Information Exchange; a 7-bit plus parity bit code.

ARPA: Advanced Research Projects Agency, now known as DARPA.

ASK: amplitude-shift keying; a digital modulation technique.

AT: access tandem; exchange area gateway switch.

ATC: adaptive transform coding; a speech coding technique.

ATTIX: AT&T interexchange carrier.

Audio conference: service in which teleconference participants communicate by voice only.

Augmented audio conference: service in which teleconference participants communicate by voice and provide graphical or textual information through additional, special equipment.

Automated office: a business environment that incorporates voice, data, and video services to improve the productivity of persons performing office-related tasks.

AWG: American Wire Gauge.

Bandwidth: the range of frequencies that just contains the energy associated with a complex signal.

Basic services: the collection, transportation, and delivery of traffic on behalf of others wherein the information delivered exactly matches the information collected. Although the signal may be changed on its journey through the network in order to facilitate passage, the message content is not affected.

BCD: binary coded decimal.

Bell operating company: one of 22 local service telephone companies, formerly components of the Bell System, now grouped into 7 regional operating companies (ROCs).

BER: bit-error rate; measures degradation of data signal.

Binary codes: line codes that employ the substitution of longer, but symmetrical blocks for blocks in the information signal, in order to maintain balance between ones and zeros.

Binary-coded decimal: a number representation in which successive digits in a base 10 number are represented by their binary equivalent.

Binary number: a number in a base 2 system; usually 0 or 1.

Bipolar: a type of semiconductor device characterized by high speed. A process for making semiconductor devices.

Bipolar codes: line codes that use a three-level code to represent the information signal; usually +1, 0, and −1.

BOC: Bell operating company.

BORSCHT: acronym for battery feed, overvoltage protection, ringing, supervision, coding and filtering, hybrid, and testing.

Bridged tap: an outside plant connection technique that results in an unused twisted pair being left connected to an active customer loop.

Broadcast videotex: teletext; an information service that is carried in a portion of the television broadcasting signal.

BT: bridged tap.

Business services: components of a medium that respond to the needs of commerce and industry; primarily intercommunications.

Call processing: a program that directly controls the switch matrix and makes connections between switch ports. Includes supervision and execution.

Capital recovery: replacement of cash spent by company to purchase equipment for use in business. The most common technique is an annual charge against revenues which repays investment (with interest) over a specific period.

Carrier: a company that transports messages (message traffic) between customers. May be regulated or unregulated.

CATV: cable television.

C band: transmission frequencies between 4 and 8 GHz.

CCIS: common channel interoffice signaling.

CCITT: Comité Consultatif International Télégraphique et Téléphonique.

ccs: one hundred (centum) call seconds; a measure of telephone traffic.

CCS: common channel signaling

CDMA: code-division multiple access; a spread-spectrum technique.

Cellular mobile radio: a technique of providing telecommunication to mobile customers that employs frequency reuse in an arrangement of geographic cells.

Centrex: service provided from the end office that implements features equivalent to those associated with a PBX.

Channel: the smallest segment of a circuit which provides transport for a single telecommunication service, such as a voice channel.

Channel bank: equipment that multiplexes a group of channels into a higher-frequency band, and demultiplexes them.

CHILL: CCITT high-level language for programming electronic switches.

Chrominance: TV signal component proportional to both the hue and saturation of each element. The phase of the chrominance signal represents the color, or hue, and its amplitude represents the boldness of the color.

Circuit: a pair of channels that provide bidirectional telecommunication.

Circuit switch: equipment that establishes a physical path between input and output ports for the duration of information exchange between customers connected to the ports.

CMI: coded mark-inversion; a bipolar line code. Symbols are represented by 01 and 10.

CMOS: complementary metal–oxide–semiconductor; a type of semiconductor device characterized by low power requirements. A process for making semiconductor devices.

CMR: cellular mobile radio.

Code-division multiple access: a spread-spectrum technique in which signals in the same radiofrequency band employ separate spreading codes so that they may be differentiated.

Commentary quality speech: high-quality voice signal suitable for broadcasting and reproduction. Natural and intelligible.

Common channel signaling: use of an additional channel, separate from voice channels, for signaling. One channel will generally serve many voice channels.

Communication: the act of imparting or exchanging information, knowledge, and ideas.

Communication quality speech: voice signal that can be understood by trained communicators. Not particularly intelligible to average persons.

Companding: compression and expansion of voice signal so as to accommodate loud and soft talkers in same equipment. Companding may use μ-law or A-law relationships.

Compiler: a software product installed in the host machine that converts high-level language statements into digital code to be used in the target machine.

Computer conference: service in which teleconference participants exchange messages through a memory associated with a computer to which all have access.

Concentration: a technique that makes a fixed output capacity available to an input community which, if all require channels simultaneously, cannot be served.

Conference bridge: a device that connects several telephone channels together so that several users may talk as a group. Connections can be made by the conference initiator, or by the participants (see *Meet-me bridge*).

Congestion: a network condition in which the offered traffic exceeds the amount of traffic which can be handled.

CPFSK: continuous-phase frequency-shift keying; a digital modulation technique.

CREG: concentrated range extension with gain.

Critical angle (of incidence): the angle of incidence of an optical ray on an optical interface at which diffraction ceases and total internal reflection begins.

CRT: cathode-ray tube.

CSDC: circuit-switched digital capability.

CSMA: carrier sense multiple access; a packet protocol.

CSMA/CD: carrier sense multiple access with collision detection; a packet protocol.

CPU: central processing unit.

D/A: digital-to-analog conversion.

DACS: digital access and cross-connect system.

DAMA: demand-assigned multiple access.

Data abstraction: action of defining additional data types which consist of related collections of information applicable to one or more modules so that access is localized and controlled.

DARPA: Defense Advanced Research Projects Agency, a unit of the Department of Defense.

DARPANET: a packet-switched data communication network supported by DARPA.

Datagram (mode): manner of operating a packet network in which packets transit the network over any available connection.

DAV: data-above-voice.

dB: decibel; a convenient measure of power and voltage ratios.

DBS: direct broadcast satellite.

DCE: data communications equipment.

DCTE: digital communications terminating equipment.

DDCMP: digital data communications message protocol; a DECNET protocol.

Deadlock: a condition in a contention network in which all servers are waiting for instructions from other servers before they can act. The result is that none can process messages.

DEC: Digital Equipment Corporation.

Decibel: a dimensionless quantity representing 10 times the logarithm (to base 10) of the ratio of two powers; also 20 times the logarithm of the ratio of two voltages.

DECNET: network employing a digital network architecture interconnecting DEC products.

Delta modulation: a differential coding technique in which the sampling rate is many times the Nyquist rate. The difference between samples is coded as $+1$ or -1 by a 1-bit quantizer.

Demultiplex: separate a multiplexed stream into individual component channels.

DEMSNET: digital electronic message network.

Differential coding: a technique that exploits the correlation between adjacent samples to reduce the transmitted bit rate.

Differential PCM: a differential coding technique in which the difference quantizer employs two, or more, levels.

Digital network architecture: an arrangement that organizes functions and provides protocols for all levels of interaction in a data communications network comprising DEC and DEC-compatible equipment.

Digital signal: a signal that assumes discrete states only. The most popular class of digital signals are binary valued, that is, they exist in two states. Changes of state occur instantly, and the rate of change is infinite.

Digital space switch: a switch employing highway interchangers.

Digital time switch: a switch employing time slot interchangers.

Diversity: employment of redundant frequencies, antenna locations, time slots, or polarizations to improve the probability of reliable transmission.

DLC: digital line carrier.

DM: delta modulation; a speech coding technique.

DNA: digital network architecture for DECNETs.

DNHR: dynamic nonhierarchical routing.

Downstream: direction of the flow of signals from the service provider to the customer.

DPCM: differential pulse code modulation; a speech coding technique.

DPSK: differential phase-shift keying; a digital modulation technique.

DRCS: dynamically redefinable character set; for information display.

DSB-AM: double-sideband amplitude modulation.

DSP: digital signal processor; an integrated circuit chip.

DSI: digital speech interpolation.

DTE: data terminal equipment.

DTMF: dual-tone multifrequency (signaling).

DTS: digital termination service for bypass of local telephone company.

Duplex (transmission): capability to operate in two directions at the same time.

DUV: data-under-voice.

EAX: electronic automatic exchange.

EBCDIC: Extended Binary-Coded-Decimal Interchange Code; an 8-bit code.

Echo control: management of echo effects by introduction of cancellation, suppression, or channel separation techniques.

Economy of scale: saving achieved in cost of individual product or service by minimizing the number of different products so as to maximize the manufacturing volume of each, or by operating a common system so as to increase the number of messages using the same facilities.

EIA: Electronic Industries Association.

Electronic blackboard: equipment that provides one-way information space in which users may draw or write items which are visible to remote participants.

Electronic mail: a medium that facilitates the non-real-time exchange of text, voice, and graphic messages between persons.

End office: a telephone facility that provides first-level switching. It is the interface node between the customer loop and the exchange-area access tandem switch.

End-to-end communication: information exchange between users without regard to how the messages move from origin to destination.

Enhanced services: the collection, transportation, and delivery of traffic on behalf of others wherein the information is processed at some node or termination in response to its content.

Entrance link: connection between multiplexer and radio. May be a few feet or several miles.

EO: end office.

EOT: end-office terminal.

Erlang: the ratio of the time during which a circuit is occupied to the time for which the circuit is available to be occupied.

Erlang B formula: expression for computing grade of service on the assumption that blocked requests are lost to the system.

Erlang C formula: expression for computing grade of service on the assumption that blocked requests are eventually served.

ES: expert system.

ESS: electronic switching system.

Exchange area: geographic division that encompasses a community with common social, economic, and other interests, and is served by one or more contiguous end-office serving areas.

Exchange-area access tandem switch: switch connected to end offices in an exchange area that provides inter-end-office switching and access to interexchange networks.

Expert system: a software system that solves problems in limited, well-defined areas, using incomplete or uncertain information.

FDM: frequency-division multiplex.

FDMA: frequency-division multiple access.

FET: field-effect transistor.

FEXT: far-end cross talk.

Flow control: a function allowing a receiving entity to control the transfer of data over a connection, so as to prevent overload or achieve other objectives.

FM: frequency modulation.

Fourier series: set of harmonically related sinusoidal functions that can be made to represent any arbitrary, repetitive signal.

Freeze-frame video: equipment that extracts a single frame from a stream of television pictures formed at normal speed and transmits it over telephone lines at a slow speed to other locations where the signal is used to reconstruct a still picture.

Freezeout: clipping of the front end of a talkspurt due to the lack of facilities to process the signal.

Frequency-division multiplex: a technique that employs equally spaced carriers to produce a composite signal of many channels which can be transported over a single transmission path.

Frequency modulation: the frequency of the carrier wave is varied (modulated) by the signal wave.

Frequency reuse: ability to use the same frequency in several locations by shaping the antenna radiation characteristics, reducing transmitted power, employing different polarizations, or employing different time slots, so that the transmissions do not interfere with one another.

FSK: frequency-shift keying; a digital modulation technique.

Full-duplex (transmission): capability to support transmission in both directions simultaneously.

Fundamental term: the lowest-frequency component in a Fourier series.

GaAs: gallium arsenide.

Gateway (switch): terminating point of major connection between separate networks.

GHz: gigahertz; one billion hertz.

GHz km: product of usable fiber bandwidth in GHz and length of fiber in kilometers.

Grade of service: a quantity describing the level of service provided in terms of the delay encountered due to contention for network functions.

Graded index fiber: optical fiber in which the refractive index of the core varies parabolically across the radius (maximum in middle).

Graphics mail: service using facsimile machines to exchange graphical material between persons.

Half-duplex (transmission): capability to support transmission in either direction, but not simultaneously.

Harmonic terms: the higher-frequency terms in a Fourier series. They are multiples of the fundamental frequency.

HDLC: high-level data link control.

Header: the part of a message that contains information used for the identification and control of routing and delivery.

Hertz: unit of frequency; one cycle per second.

HI: highway interchanger.

High-level language: a programming language that contains control statements understood by persons.

Highway interchanger: a digital switching element that moves information from one bus (highway) to another by interconnecting them for the duration of the appropriate timeslots.

Holding time: length of time a circuit remains connected.

Homing arrangements: instructions that define the alternative routes to be used for traffic which overflows the links directly serving the requested connection.

Host machine: computer in which program is installed and executed.

HSSDS: high-speed switched digital service; a 3-Mb/s connection.

Hundred call seconds: one hundred call seconds is the traffic due to one call of 100 s duration, 100 calls of 1 s duration, or any combination in between.

Hz: hertz.

IBM: International Business Machines Corporation.

IC: integrated circuit.

IDN: integrated digital network.

In-band signaling: use of signaling tones (dual-tone multifrequency, for instance) which fall within voice band.

Independent radio carrier: a company providing radio paging and/or mobile telephone service.

InGaAs: indium gallium arsenide.

InGaAsP: indium gallium arsenide phosphide.

InP: indium phosphide.

Integrated circuit: a device constructed on a small square of silicon which contains passive components (such as resistors and capacitors), and active components (such as transistors), interconnected to perform a circuit function.

Integrated digital network: a network in which connections established by digital switches are used to transport digital signals.

Integrated services digital network: integrated digital network that provides transport for voice and non-voice services.

Interactive broadcast television: television supplied over broadcast television facilities with response by telephone (or other network).

Interactive cable television: television supplied over cable television facilities with response by return circuits arranged over cable television facilities.

Interactive television: a medium that serves many persons simultaneously, providing a selection of video programming to them, and carrying responses in the return direction.

Intercommunications services: components of a medium that allow the exchange of voice, data, or video messages between two customers, or among a small number of customers in conference.

Interexchange carrier: a company that transports messages between exchange areas (in different local access and transport areas). Connection is normally made at an exchange-area access tandem switch.

Intersymbol interference: degradation of signal pulse train due to extension of tail of one pulse to overlap leading edge of next pulse.

Ionosphere: ionized layer in earth's atmosphere produced by the impact of the sun's rays on the gas molecules of the upper atmosphere.

IRC: independent radio carrier.

ISDN: integrated services digital network.

ISO: International Organization for Standardization.

ISO interconnection model: an architectural concept to aid the definition of protocols necessary to interconnect and facilitate communication between data users.

Isotropic antenna: an antenna that transmits or receives signals equally well in all directions.

JJ: Josephson junction; a cryogenic digital device.

Ka band: transmission frequencies between 26.5 and 40 GHz.

Keying: digital modulation.

Ku band: transmission frequencies between 12.5 and 18 GHz.

LADT: local area data transport.

LAN: local area network.

LATA: local access and transport area.

Layer: an element in the architecture of the ISO model. A group of related entities between upper and lower logical boundaries. Often referred to as a level.

LD: laser diode.

LED: light-emitting diode.

Level: alternative name for layer in ISO model or a quantization point in analog-to-digital conversion and digital coding.

Linear prediction coefficients: quantities used to define a linear prediction model which estimates speech samples.

Line codes: coding employed in digital systems to reduce low-frequency and zero-frequency energy in the signal bandwidth.

Line of sight: a direct ray from receiver to transmitter.

Loading: the addition of inductance to a wire pair connection to improve signal transmission.

Local access and transport area: consists of one or more contiguous exchange areas which form a local calling area.

Local area network: a packet telecommunication facility that employs a common bus to interconnect terminals and data processing equipment.

Logical connection: an abstract connection said to exist between two entities to simplify understanding of data transfer. The physical implementation is hidden from the user.

LSI: large-scale integration of circuits on silicon chips.

Luminance: TV signal component proportional to the brightness of each element of the scene.

Majority carrier: electrons or holes that provide the greater charge transport in a device (see *Minority carrier*).

Market pull: the attraction of a specific set of customer needs for a particular supplier of products or services.

Mass communication services: components of a medium that provide voice, data, or video messages to any number of customers (usually in a limited area).

Matrix: that part of a stored program-controlled switch which connects the two sets of customer lines together so that information can be exchanged.

Medium: an agency or means that facilitates action at a distance.

Meet-me bridge: a conference bridge used by groups of persons who call the bridge operator at a prearranged time and are connected to each other so that they may talk as a group.

Message of the medium: the change of scale or pace or pattern introduced into human affairs.

MF: multifrequency (signaling).

MFJ: Modified Final Judgment.

Micron: one-millionth of a meter.

Microsecond: one-millionth of a second.

Microwave frequencies: frequencies close to and above 1 GHz.

Milliwatt: a quantity of power equal to one-thousandth of a watt.

Minority carrier: electrons or holes that provide the lesser charge transport in a device (see *Majority carrier*).

MMIC: monolithic microwave integrated circuit.

MML: man–machine language developed by CCITT.

MOCVD: modified chemical vapor deposition; an optical fiber-making process.

Modem: contraction of modulation and demodulation; a DCE employed to condition data signals for transmission over voice channels.

Modified Final Judgment: settlement of antitrust suit brought by Department of Justice against American Telephone and Telegraph Company and Western Electric Company based on the contention that AT&T had used its control of local exchange facilities to frustrate competition in supply of telecommunication services.

Modulation: the action of mixing two signals of different frequencies to produce a modulated signal. The higher-frequency signal is known as the carrier.

MOS: metal–oxide–semiconductor; a type of semiconductor device. A process for making semiconductor devices.

MSK: minimum shift keying; a digital modulation technique.

MTSO: mobile telephone switching office.

μ-law companding: a modified logarithmic relationship used to shape compression and expansion characteristics of equipment to accommodate the levels produced by loud and soft talkers.

Multiplexing: a signal processing technique for combining many channels into a complex signal which can be transported over a single transmission path.

μs: microsecond.

mW: milliwatt.

NABTS: North American Broadcast Teletext Standard.

Nanosecond: one-billionth of a second.

NCP: network control protocol; employed in ARPANET.

NEXT: near-end cross talk.

Noise: interfering signals that mask or distort the signal carrying information of present interest.

Nonblocking (network): term used to describe a facility that can handle the maximum traffic load likely to be presented.

ns: nanosecond.

NSP: network systems protocol employed in DNA.

NTSC: National Television Standards Committee.

OC: operations center for network management.

OCC: other common carrier.

Office automation: the use of technology to improve the effectiveness of persons performing office-based tasks.

1/f noise: an approximation to the noise observed in very wideband devices. It contains an increasing amount of energy at lower frequencies.

On-line software: a unique package of code tailored to a specific end office or other site, generated from an office-independent program, which includes the capability of performing all of the functions of the product line and office-dependent data.

Open systems architecture: an arrangement that organizes functions in a data communications network providing protocols for all levels of interaction. Basis of OSI model.

OSA: open systems architecture.

OSI: open systems interconnection (model); a model describing data protocols.

OSIE: open systems interconnection environment.

OSP: outside plant.

Other common carrier: company providing interexchange transport in competition with ATTIX and others.

Out-of-band signaling: use of tone between edge of voice band (3400 Hz) and edge of channel (4000 Hz) for signaling.

Outside plant: the sum of telephone facilities connecting each customer to the serving end office.

OVD: outside vapor deposition: an optical fiber making process.

Packet: a grouping of bits that is collected, transported, and delivered as a unit. Standard packets consist of fields assigned to flags, control bits, address, message, and checksums.

Packet switch: equipment that contains bus-interconnecting ports on which packets contend for transit.

PAL: color coding scheme for TV signals employed in Western Europe and much of the world.

Partial response coding: a binary line code in which ones and zeros are encoded as sets of two pulses (+1, −1 or −1, +1). The second pulse of one symbol occurs at the same time as the first pulse of the next symbol. The result is a series of +1, 0, and −1 levels.

Parity: an extra binary signal (1 or 0) is added to each word to make the total number of 1's or 0's even or odd. This technique is used for error checking.

PBX: private branch exchange.

PCM: pulse code modulation.

PDI: picture description instructions for information display.

Phase modulation: the phase of the carrier wave is modulated by the signal wave.

Phonemes: basic sounds which, when strung together, form speech.

Physical connection: an actual connection consisting of physical hardware that carries specific signals.

PIN: photodiode made from *n*- and *p*-type materials separated by intrinsic silicon to extend the depletion region.

PLP: presentation level protocol for videotex.

PM: phase modulation.

Poisson formula: expression for computing grade of service on the assumption that some blocked requests are lost and some are served eventually.

Polarization: a property of electromagnetic waves related to the direction of the electric vector.

Presentation level protocol: de facto North American standard for videotex.

Principle of superposition: the response of a linear system to a complex repetitive signal is the same as the sum of its responses to the sinusoidal signals that make up the Fourier series which represents the complex signal.

Procedure: a software operation whose execution causes change to occur.

Protocol: a set of procedures that facilitates communication between terminals and machines through a single network, or over several networks.

PSK: phase-shift keying; a digital modulation technique.

Pulse code modulation: a digital signal containing strings of bits that represent successive samples of an analog signal. As 8-bit pulse code modulation, that is, each sample is represented by an 8-bit word, it is the coding used by administrations around the world to encode digital voice.

QAM: quadrature amplitude modulation; a digital modulation technique.

QPR: quadrature partial response; a digital modulation technique.

QPSK: quaternary PSK; a digital modulation technique.

Quadrature amplitude modulation: a digital modulation technique employing equal-rate binary streams.

Quadrature partial response: a digital modulation technique employing two equal rate three-level duobinary streams.

Quality of service: deals with overall performance of connections, once made. Includes factors that distort, destroy, or introduce error into the message.

Quantizing error: difference between signal levels assigned in a digital transformation and the actual value of the samples of the original signals.

Raised cosine pulse: a spectrum minimizing the unidirectional pulse shaped as a full cosine function.

RAM: random access memory.

Real-time media: combinations of services that deliver a continuous stream of information as it is produced.

Record media: combinations of services that deliver a limited amount of information in a finite time, and preserve it for study.

Regional operating company: holding company consisting of Bell operating companies (BOCs) formed as a result of the Modified Final Judgement. The 22 BOCs are separated from AT&T and organized in 7 Regional Operating Companies.

Regulated (carrier): company providing service in a franchised area. Charges are supervised by a public commission.

Residence services: components of a medium that respond to the needs of the household. Primarily mass communications.

Richness of medium: quality that affects outcome of teleconference. Richest conference medium is face-to-face without electronic mediation. Rich medium is two-way color video. Less rich medium is telephone augmented by graphics equipment. Least rich medium is text mail.

ROC: regional operating company.

ROM: read-only memory.

Routing: technique used to find the most efficient path between a source and destination.

Run length coding: a technique in which strings of the same symbol (or symbol combination) are represented by the symbol and the number of times it is repeated.

S: space, as in digital space switch.

SAC: serving area concept.

SAI: serving area interface.

Sampling theorem: by sampling an analog signal at a rate equal to twice the frequency of the highest component to be preserved, the signal can be reconstructed to contain this frequency, but none higher.

SBC: subband coding.

SDL: Specification and Description Language developed by CCITT.

SDLC: Synchronous Data Link Control; a protocol employed in an SNA network.

SECAM: color coding scheme for TV signals employed in France, USSR, and Eastern Europe.

Service: in ISO model, a service is the set of capabilities a layer offers to the user in the next higher layer.

Session: a time-limited logical connection between computer systems.

Shot noise: unwanted signals produced by the discrete nature of electronic motion in semiconducting materials.

Signaling: the action of providing information to various facilities to establish and take down calls, and report on the progress of a call.

Silence detection: identification of periods when speech is not present; detection of cessation and onset of vocalization (talkspurt).

Single-mode fiber: optical fiber in which the core is a few microns in diameter and the refractive index is constant. Only one optical mode is transported along the fiber.

SM: statistical multiplex.

SMSA: Standard Metropolitan Statistical Area.

SNA: Systems Network Architecture for IBM computer networks.

Software engineering: provides rules and guidance for the development of large software products using a team of persons.

Space division: a technique for providing separate channels by assigning a physical path to each channel.

SPC: stored program controlled (switch).

Speech quality: assignment of a level that identifies naturalness and intelligibility of communicated speech. Four levels have met with a measure of acceptance: commentary, toll, communication, and synthetic.

Spread-spectrum: a radio modulation technique that employs a very wideband signal created through the use of an independent code (spreading function). The composite signal has a very much higher bit rate than the original signal.

SSB-AM: single-sideband amplitude modulation.

SSB-SC: single-sideband suppressed carrier modulation.

Statistical multiplex: a technique that combines voice or data signals in a higher bit rate stream by sharing the time available among the talkspurts or packets from the users.

Step-index fiber: optical fiber in which the refractive index of the core is constant and greater than the cladding.

Store-and-forward: a feature in which a message is collected in memory and transmitted to its destination at a later time. Storage may be for a few milliseconds, or for much longer times depending on the nature of the service provided.

Stored program controlled switch: switching equipment that incorporates a digital processor to direct switching actions under the control of software programs stored in the processor.

Structured message processing: development of further actions to be performed by office automation facilities consequent on the receipt of a message and the analysis of its contents.

Subband coding: coding in which the signal is split into several frequency bands and each band is coded separately.

Supervision: the action of monitoring customer lines and trunks to detect requests for actions.

Switch: equipment that selectively establishes and releases connections between transmission means so as to allow information to flow between any two points. Switches allow reuse of facilities and concentrate traffic.

Synchronization: the manner in which entities ensure the exchange of digital information is coordinated; an action that maintains one repetitive (digital) process in step with another.

Synthetic quality speech: the quality of a voice signal reconstructed from information derived from modeling the human voice production mechanism.

Systems Network Architecture: an arrangement that organizes functions in a data communications network providing protocols for all levels of interaction.

System-on-a-chip: an integrated circuit that consists of a large enough number (several tens of thousands) of elements to replicate the capability of a contemporary electronic system on a single chip. Thus, PBX-on-a-chip, etc.

T: time, as in digital time switch.

Talkspurt: a connected set of phonemes that make up a word or phrase; a period of time in which a person utters sounds.

Target machine: computer on which program being written is to run, hence, target code, or more usually, object code.

Tariff: information filed with regulatory authority that describes performance of, and charges for, a specific capability offered by a regulated carrier.

TASI: time-assignment speech interpolation.

TCM: time-compression multiplex.

TDM: time-division multiplex.

TDMA: time-division multiple access.

Technology push: the stimulation of customer demand by the demonstration of the application of new products or services.

Telecommunication: communication at a distance.

Telecommunication facilities: combinations of equipment that collect, transmit, and deliver information by electrical, electronic, and optical means.

Telecommunication media: combinations of services that collect, transport, and deliver messages.

Telecommunication services: components of a telecommunication medium that respond to specific customer needs.

Teleconference: electronic communication between three or more persons at two or more separated locations.

Teletex: service in which text messages are automatically exchanged between electronic storage typewriters or communicating word processors.

Teletext: a one-way information system that employs unused lines in the broadcast television signal to transport data to modified television receivers without interfering with the normal program.

Telex: a service that automatically transfers messages over slow data rate channels between teletypewriters.

Text mail: service that allows the exchange of printed messages between persons.

Thermal noise: unwanted signals due to the random motion of electrons in the signal-carrying medium produced by thermal energy.

3DI: three-dimensional integration of semiconductor circuits.

Three-dimensional integration: a technique in which individual integrated circuits are fabricated one on top of the other to produce a three-dimensional structure.

THz: terahertz; one million, million hertz.

Time-compression multiplex: a technique to establish two-way digital transmission over a pair of wires. The equipment at the two ends of the connection share the circuit. They alternate in sending bursts of information.

Time division: a technique for providing separate channels by assigning a unique succession of time slots to each channel.

Time-division multiplex: a technique that combines digital signals in higher-rate channels by sharing the time available among them.

Time slot interchanger: a digital switching element that disassembles and reassembles bit streams in different sequences by storing (delaying) the appropriate bits.

Time warping: the action of distorting the time base of speech utterances to better fit standard templates.

TMUX: transmultiplexer.

Toll office: a switching office that routes calls in the interexchange network.

Toll quality speech: easily understood voice signal suitable for carrying conversations between average persons located anywhere in the world. Natural, intelligible, but obviously limited by characteristics of telephone facilities.

Top-down design: technique for design of large software products that starts with a simple statement of the objective and decomposes the work through a hierarchical structure to single functions which are small enough to be programmed, tested, and integrated by a single analyst.

Traffic: signifies the amount of information flow through a network; measured in erlangs or ccs.

Transistor: amplifying device constructed from very pure crystal of silicon. The discovery and development of this device launched the solid-state age.

Transmultiplexer: equipment that converts frequency-division multiplex streams of analog voice channels to time-division multiplex streams of digital voice channels (and vice versa) without demultiplexing.

Transparency: a concept in which information flow does not appear to be affected by changes in the transport mechanism. While the changes exist, they are hidden from the users.

Transport service: service solely concerned with moving (transporting) messages (traffic) from customer to customer.

TSI: time slot interchanger.

ULSI: ultralarge-scale integration of semiconductor devices.

Unregulated (carrier): company providing message services in a competitive environment in accordance with demand.

Upstream: the direction of flow of signals from customer to central control and origination point of programming.

VAD: vapor-phase axial deposition; an optical fiber making process.

Vertical interval: time required by electron beam in a TV camera tube to move from bottom of target to top. No picture signal is generated. Instead a number of empty horizontal sweeps produce lines that can be used for other messages (such as teletext).

Videoconference: service in which teleconference participants communicate by means of video links.

Videotex: a medium that facilitates information retrieval by many persons from remote databases. Implemented as wired videotex and broadcast videotex (also known as viewdata and teletext, respectively).

Viewdata: a two-way interactive information system that uses a video display (often a television receiver), local processing, and a remote database accessed through the telephone network.

Virtual circuit (mode): a method of operating a packet network in which a stream of packets generated at one terminal transit the network along the same route to their destination.

Virtual connection: an abstract connection between peer entities. A logical connection.

VLSI: very large scale integration of semiconductor devices.

Vocoding: a technique of voice encoding that emulates the vocal tract mechanisms. Coding is accomplished by describing the sound production parameters and decoding is done by using them in circuits that simulate the voice production capability of the vocal tract.

Voice mail: service that records the words spoken by sender, stores them, and delivers the sound of the sender's voice when the recipient calls for messages.

Von Neumann machine: a processor employing sequential processing.

VSB-AM: vestigial sideband amplitude modulation.

Wafer-scale integration: a technique that places different circuits on the same silicon wafer and interconnects them to make an entire system on the wafer.

WAL2: a bipolar line code. Symbols are represented by 0110 and 1001.

Watt: a unit of power.

Wavelength: the distance traveled by energy of a specific frequency during one cycle.

WDM: wavelength-division multiplexing.

White noise: an approximation to the sum of thermal and shot noise. White noise has a constant average noise energy spectrum across the frequency band of interest.

Windows: portions of the frequency spectrum above 10 GHz that are relatively free from absorption loss due to water, oxygen, and other elements.

Wired household: a household that includes a pervasive communication–information system.

Wired videotex: viewdata.

Wireline carrier: local service telephone company.

WSI: wafer-scale integration.

X.25: CCITT protocol that defines a universal interface between data terminals of varying sophistication and public packet-switched network.

X.75: CCITT protocol that defines the interconnection of data networks.

Zero-frequency term: the constant term in a Fourier series.

Index